A. M. Halsted

Artificial Incubation and Incubators

A. M. Halsted

Artificial Incubation and Incubators

ISBN/EAN: 9783337662240

Printed in Europe, USA, Canada, Australia, Japan

Cover: Foto ©berggeist007 / pixelio.de

More available books at **www.hansebooks.com**

ARTIFICIAL INCUBATION

AND

INCUBATORS.

ILLUSTRATED.

A TREATISE ON

RAISING POULTRY BY ARTIFICIAL MEANS,

WITH

DESCRIPTIONS AND ILLUSTRATIONS OF EVERY
INCUBATOR OR HATCHING APPARATUS
WORTHY OF NOTICE,

IN ANCIENT OR MODERN TIMES.

By A. M. HALSTED,

RYE, N. Y.

SECOND EDITION.

REVISED AND ENLARGED.

NEW YORK.
ALBERT METZ & Co., Printers, 60 John Street.

1883.

Dedicated to that most genial and whole-souled body of men—
The Poultry Fraternity.

CONTENTS.

INTRODUCTION.

When, nearly two years ago, the necessity for a work like this became apparent, and the author was asked by numerous friends to prepare and publish one, it was with reluctance that he accepted the task; for, although he had had his share in shaping, and contributing to the poultry literature of the country for the past ten years, he felt that to undertake a work such as was needed and demanded by the fanciers of the country required no small amount of study and research, and he doubted his ability to complete it in as thorough and comprehensive a manner as he knew it should be done.

The assurances of friends, however, and the seeming success of other writers, who, he felt assured, had not a tithe of the experience which necessity had compelled him to acquire, has encouraged him to persevere in the completion of the work. As first contemplated, he did not intend it should be over half the number of pages, but the subject grew on his hands, Incubators multiplied in numbers, and were he to mention all inventions of this nature which have received recognition at the U. S. Patent Office, he would be obliged to add another score to the number already described.

In gathering information, he wishes to acknowledge indebtedness to Balfour's Embryology, to Wright's Practical Poultry Keeper, and also to the columns of the various poultry papers.

That there may be no misunderstanding regarding the notices of Incubators which appear in Chapter X, he would state here that every manufacturer, whose address was known, was notified by letter of his intention to devote a chapter to such notices, and given an opportunity to write what he wished to have said about his invention. Less than half a dozen accepted or replied; their communications appear in their own words, and over their own signatures and addresses. All information regarding other machines has been compiled from such sources as were available; in some cases he had personal knowledge of the machines, and of their strong and weak points; in others the information has been derived from circulars and books, and, in some few instances, from verbal communications from those who had used the Incubators. Of course, under such circumstances, the inventor's name should not, and does not, appear.

His object has been not to compile a collection of circulars and advertisements, but to make the work as complete as possible, that it may be a book of reference to searchers after knowledge on this subject, and a medium of instruction to those who wish to try hatching and rearing poultry by artificial means.

If he has succeeded in this effort, of which his readers must be the judges, he is content.

A. M. HALSTED.

RYE, N. Y., JULY 1ST, 1880.

PREFACE TO SECOND EDITION.

The success attending the First Edition, and the very complimentary notices received from its readers in all parts of the country, encourage the author to revise and enlarge the work, before presenting the Second Edition to the public.

This he has done by adding a number of pages to several of the chapters, entering more fully into the details of the subjects. The chapter on Incubators also has been enlarged by the illustration and description of a number of Machines, which have been brought before the public since the issue of the last edition. A new chapter is added to the work, in response to numerous requests for such information, treating of Houses, Yards, Location, etc., and giving illustrations pertaining to the subject.

Again submitting his work to an appreciative public, he awaits their verdict.

A. M. HALSTED.

RYE, N. Y., *May 1st*, 1883.

ARTIFICIAL INCUBATION AND INCUBATORS.

CHAPTER I.

GENERAL REMARKS.

It appears to be a prevalent idea that while poultry in small flocks will pay a good profit to the keeper, in larger numbers the losses more than counterbalance the gains.

This opinion is well supported by facts, for the record of failures in the attempts to raise and market poultry in large numbers —in other words, to make a business of it—are numerous, while the instances of success are very rare.

Yet, that it can be made successful is beyond a doubt. In nearly every case of failure which the writer has investigated, there has been found, first, no practical acquaintance with the business; and second, it has been made a secondary matter: some other occupation or employment occupying the most of the thoughts and time of the owner, the poultry business being expected to run itself, and requiring only a casual supervision of the attendant morning and night. As well might one expect to transact a banking business or any other mercantile pursuit, and be absent from his office or store three-fourths of the time.

The largest enterprises often have their origin in very small beginnings; the projectors of such know and appreciate the folly of attempting to conduct any large business upon the same capital and with the same clerical help as in its incipient stages; yet the majority of those who go into the poultry business appear to think that it is an exceptional occupation, requiring nothing but some houses, some fowls, some feed, and a boy or man to supply the one to the other. Herein is the fatal mistake. Apply the same capital, the same business tact and enterprise, give it one's whole time and attention, and success is quite as certain as in any other pursuit. The same watchfulness and attention to business is as requisite here as in the dry goods or grocery trade. Supplies must be purchased when and where is most economical, taking every advantage of the market. It is poor policy to buy any article that can be produced on the place at a less cost; and *vice versa*, it shows lack of business tact to raise corn, grain, or anything else, which can be purchased cheaper in the market. So, also, regarding investments in buildings, stock, etc. If a house costing one hundred dollars will answer the purpose intended, and can be built for that sum in a substantial manner, every dollar expended above that amount is virtually wasted.

Another view, however, may be taken of this subject; houses may be made so cheaply that constant repairing and, in fact, almost rebuilding, will be necessary every year. This is as unwise an investment as the former, and equally to be condemned.

In the purchase of stock, also, economy may be carried too far. A lot of fowls, averaging three or four pounds each, costing fifty cents apiece, will be very dear breeding stock if selected in preference to others of twice that weight, and costing double the price. The first will dwarf their progeny, while the last will produce their like; in a season's breeding, where a thousand or more chickens are raised, the difference in the profits would be an item of no small amount.

Taking only one thousand chickens as the amount produced, and allowing a difference of but two pounds each in gain for the large fowls over the smaller ones, we have an increase of two thousand pounds of marketable flesh, which, at ten cents per pound, is an item of two hundred dollars to the profit side of the account.

The small profits must be closely looked after. The droppings carefully saved and sold, or used on the land. The sale of feathers in a large establishment make quite an item. The saving of food by the use of improved feeding utensils, and the saving of time by using, as far as economical, labor saving tools and conveniences about the houses and yards. It is by attention to the small minutiæ of any business that success is assured. "Take care of the pence, and the pounds will take care of themselves."

Help enough must be kept to ensure the necessary care of the stock, but at the same time the work may be so systematized that three pair of hands can do the work which heretofore has required four.

The arrangement of the buildings and their location are of considerable importance; all necessary conveniences should be added to facilitate the work. The feeding system so arranged as to require the least possible time in the distribution of food. The fattening arrangements constructed on the most approved plans, consistent with the available capital of the builder; the saving of time and feed will be found an item of much importance.

Marketing is another point to be considered, where it is proposed to raise poultry for sale. The distance to a good market, and the cost of sending the products there, will sometimes absorb so large a share of the profits, that the business is practically unprofitable. It is not only necessary that there be a good market within reach, but the marketing facilities—(*i.e.*) the means of reaching the place quickly and frequently, and at a moderate expense, be also assured.

Boat routes are always the cheapest, but they have the disadvantage of being uncertain in stormy weather, and frequently discontinued during the winter and early spring. Therefore, when a place is selected on such a route, it is necessary also, to be within teaming distance of a railroad, that the market may not be shut off, during part of the year.

In marketing Broilers, a near market is almost a necessity; for to get the best returns, they should be sent in dressed; this cannot be done over a long route during the warm weather, at which time they bring the best price.

Finally, the sanitary measures must be of the best, and no relaxation of vigilance permitted in following up the enforcement of cleanliness, both in the buildings and the feeding and drinking vessels.

The houses and coops should be frequently cleaned and swept, and a liberal use of whitewash made at least twice during the season. If a little crude carbolic acid be dissolved with the wash, it will add very much to the healthfulness of the fowls. Drainage must be looked after, as a wet or damp poultry house and runs begets cholera and roup in old fowls, and rheumatism and disease in chickens.

In all departments of the enterprise, *thoroughness* is of the greatest importance. Have a complete system about the work and its execution, and see that it is fully carried out : do not be satisfied with any make-shift, but let the motto be, "*well done is twice done.*"

CHAPTER II.

ARTIFICIAL INCUBATION.

It is but a few years since that the idea of hatching eggs by other than natural means was scoffed at as the visionary notion of some over-sanguine inventor.

That it had been done, and was still a matter of actual every-day business in Egypt and some other warm countries, was well known; but there the operation was conducted so secretly that none but the regular attendants knew anything of the process.

"Eccaleobions", "Potolokians", and "Incubators" have been made and used with varying success in this country for about forty years: and in England and France for double that period. But until within the past three years, they have never given sufficiently successful results in the hands of any but their inventors to become of any use to the community at large.

Artificial Incubation, in theory, is one of the simplest things imaginable. All that is necessary is to keep the eggs at the proper temperature during the time necessary to bring forth the chickens.

In practice, however, that very necessary evenness of heat is found to be impossible, unless one of two expedients be employed. One of these is to sit by the apparatus and watch it constantly; and the other is to employ some mechanical device to make the machine control its own heat—in other words, to be automatic or self-regulating.

The positive necessity of this is obvious to every one who has attempted to hatch chickens artificially: for no matter how nicely a machine may work in a room of even temperature, when subjected to a variable temperature, it cannot retain the same heat.

If, to use an Incubator without any regulating apparatus, it is necessary to have an evenly heated room, why purchase any machine at all? A tank of water with a lamp under one end, and a shallow tray of eggs under the other, will answer every purpose. If the heat of the room is kept uniform, and the flame of the lamp of even height, the heat generated by and radiated from the tank of water will also be regular and even; but if the temperature of the room varies, the tank of water will vary with it.

As well attempt to run a steam engine without a governer as an Incubator without a regulator.

Another practical drawback in Artificial Incubation is the question of moisture.

How much, and how supplied? Theories are of no avail here. Practical experiment alone can determine the matter. Theory gives a probable reason for its necessity, but does not supply the want.

The first step the writer took in this direction was to test the question of evaporation of moisture from the egg while under artificial heat. I placed fresh eggs, selecting those of equal weight, in the Incubator, and under hens: the former were subjected to a dry heat. At the end of three days there was a discernable difference, and at seven days a very marked variation in their weights, while at fifteen days there was over half an ounce difference.

Now, to apply the theory: an egg, after having been under a hen a few days, will be observed to have a polished appearance, as if oiled, and rubbed. There is probably some oily secretion on the feathers of the fowl which, with her action in moving the eggs, produces this appearance. This coating, slight as it is, suffices to prevent rapid evaporation, and yet does not clog the pores of the shell.

Eggs placed in the Incubator always retain the same fresh appearance, and the evaporation is much more rapid; hence the necessity of supplying moisture artificially to keep the egg in as natural condition as possible. The effect of this evaporation is to harden and toughen the white membraneous lining of the shell so that when the time comes for the chicken to hatch, he cannot pick his way through, and he is as hopelessly immured as if locked in a sheet iron box.

The artificial application of moisture has been made in a number of ways: by sprinkling, which is the mode usually employed; by charging the air with moisture from pans or trays of water placed in the egg chamber; by heated pipes, which vaporize tanks of water through which they pass, and by keeping flannels or other substances saturated with water under the eggs. The first two are the only safe methods, as the others are liable to rot the eggs.

Next, and equally important, is the subject of ventilation. The writer was among the first, if not the very first, who publicly stated that ventilation was imperatively necessary in the egg chamber. A tightly closed drawer or tray, such as was first used, was always productive of bad odors. The exhalation from the eggs, which was largely of carbonic acid gas, always sank to the bottom of the egg drawer, and was fatal to the life of the chicks.

Experiment soon decided that a certain amount of ventilation was necessary to carry off this gas, as well as to cool the eggs, should the drawer get too hot; but also that more than a certain amount was unnecessary, and a waste of heating power.

Then arose the question of the proper heat and how to apply it.

In natural incubation the heat is supplied from above. This is the case in all kinds of eggs. The birds' nests are of twigs and grasses, allowing free passage of air under the eggs. Even the reptiles' eggs—alligators and turtles—lying in the sand, receive the sun heat on top; and the hen that makes her nest in the thicket, or under the brush heap, usually brings out a larger and stronger brood of chicks than one which has her nest in the most costly poultry house.

Nearly all the early efforts at Artificial Incubation were made on the principle of surrounding the eggs by heat. Reaumur, one hundred years ago, tried the process with horse manure, and achieved a partial success only. Other machines constructed on this principle resulted in virtual failures, and one after another dropped out of sight. In fact, it has been proved over and over again that eggs subjected to a uniform heat above, below, and around them, will not hatch any satisfactory percentage.

Other efforts were made with currents of hot-air, generated by lamps or stoves; these, too, were only partial successes. Heated air cannot be made to diffuse itself from any given point, laterally and evenly over any surface. To effect this, the heat must be generated and communicated to the eggs by some other medium, one which will give a uniform, steady, even temperature over all parts of any desired surface. Water seems to be the only available agency, and the most easily controlled. Confined in a vessel of suitable shape, and caused to circulate by peculiar modes of construction, it has been found to give what is required, viz: a moist, regular heat, applied from above, evenly distributed, and equally over all parts of the desired surface.

As is well known to most breeders, the germ, or life principle of the egg, always floats on the top. Under the influence of the heat applied from above, the arteries and veins expand and extend, following the inside of the white until they reach completely around and encircle the yolk. The extremities of these veins are very fine and delicate. Nature, in her wise provision for the best development of their growth, ensures the bottom of the egg, where these fine veins are, being kept rather cooler than the top. The contact of the body of the mother bird with her eggs, heating the top of the eggs and keeping that portion several degrees warmer than the bottom.

When, therefore, the heat is applied under or all around the eggs, these fine veins, instead of growing, shrivel and dry up; the yolk sack, instead of being absorbed by the growing chicken, dries fast to the shell, and about the fifteenth or sixteenth day the chick dies. Some may reach the twentieth day, and a few may and often will hatch out; but the percentage is so very small that it virtually amounts to a failure.

The proper heat has been the subject of much debate on the part of the various manufacturers of Incubators. The degree adopted varying from 90° to 106° degrees Fahrenheit. One French author and mechanician placed it at 90°. An American experimenter, E. Bayer, put it at 102°; M. Cantelo of London at 106°; M. Vallee of France at 104°, and upwards. Later experimentalists seem to have followed in the footsteps of their predecessors. One places the degree required at 106° at first and gradually decreasing to 100°. Several at 104° and decreasing. My own custom has been 101°, increasing to 104°, and a considerable number of those who at first tried the reverse system, have discarded it and now use mine.

The great diversity of opinion on this subject fifteen or more years ago, necessitated the solving of the question by actual experiment. I therefore had thermometers made of different shapes: some with flat bulbs and curved stems to place between the wings and the body; others with broad flat bowls to place under the breast of the hen; others again to test the internal heat, and still others to get the heat between and under the eggs. The result of the experiments, covering a period of several weeks, and testing above forty hens, was to establish an average heat of 103° for the three weeks. This was the contact heat at top of the eggs; the heat at the center of eggs was about 101°, and the bottom, or under the eggs, 98°; and in one case, where the nest was on the bare ground, as low as 94°.

I know that some writers state positively that the heat of the hen is greatest at first, when the sitting fever commences. A little thought will convince any unprejudiced person that this cannot be so. When the hen first begins sitting, her breast is covered with feathers, which are known to be a non-conductor of heat: these interpose between the body and the eggs, and consequently lessen the heat. As the time goes on, these feathers come out, and by the end of the three weeks the bare breast is on the eggs. Consequently the eggs get an increased heat as the time of incubation passes on. Another thing to be taken into account, is that the hen sits closer towards the last of the time than she does at first. So we have two reasons why the heat is greater at the close than at the beginning.

There are now made over thirty different styles of Incubators, some are patented, others are not: of this number there are probably half-a-dozen that will do good work in the hands of any intelligent person; perhaps half-a-dozen more which their inventors can hatch with, but with which the success of others would be extremely problematical; and the remainder are not much better than the tin pan with a lamp under one end and a tray of eggs under the other. Some of them, in fact, being perfectly worthless for the purpose, and only a fraudulent means of obtaining money from a class of people who always make cheapness their one criterion of value.

In advising the fanciers which Incubator to buy, it is a little difficult, in our position as inventor and manufacturer, to be entirely impartial; still, we shall strive in these pages to allow the claims of no one machine to be advanced in any way that may prove detrimental to another.

In selecting an Incubator—which we take for granted every breeder of any extent will be obliged to do within the next few years—we should advise, first, send for circulars of the leading sorts, those which you know are in successful operation. Examine well their claims to superiority; use your own common sense in determining questions which are in dispute between rival manufacturers; ascertain from older fanciers than yourself, which manufacturer brings to his work the ripest judgment, and fullest experience. Find which kind will best suit your own special need, and is best adapted to your building or room. Examine testimonials; we well know that many of these are supposed to be manufactured for the occasion, but when you see the names of prominent fanciers, known throughout the length and breadth of the land, appended to such, it is safe to believe that they are genuine. It is only the unknown names, in out of the way places, that we are apt to distrust.

Second, receive with the greatest caution all accounts of failures and stories of the worthlessness of rival machines. The best machines made will sometimes make failures, either wholly or partially, from one cause or another; and he who uses the mischances of a rival to bolster up his own reputation, or can find no better mode of advertising his own goods, than by running down those of his cotemporaries, had better be dealt with very cautiously, or left wholly alone.

Lastly, beware of "secret processes". They are a fraud, patent to every one who has ever made Artificial Incubation a study. We shall give in these pages all the "secrets" that are necessary to full and complete success.

The only "secrets" in the business are those said to be known by the ancients, who, it was stated, possessed the means of determining in advance of incubation, whether or no the egg was impregnated or fertile, and also the sex of the un-incubated embryo. We very much doubt if either could or can be told. There have been many theories promulgated about it, but all have been tested and found worthless.

The other so-called secrets of the Egyptians, in their practice of artificial incubation, were simply "experience," gained by the teachings of their predecessors and the necessities of their business. The testing of the heat, and the proper amount of moisture to be applied, without a thermometer, being no more difficult than the testing of the old-fashioned brick ovens by our grand-mothers, before putting in a batch of bread.

CHAPTER III.

This is a matter upon which it is difficult to go into particulars, for the forms and kinds of machines vary so much, and the directions for their care also differ so, that the advice applicable to one may be wholly unsuited to another.

I can, therefore, only generalize, or else refer more particularly, when necessary to enforce a certain idea, to the Incubator I have had most to do with—the Centennial.

The old saw, "Many men of many minds", was never more applicable than in connection with this subject. Hardly any one but has his special preference, or, perhaps we might say, adaptation to some particular pursuit or business; and many are they who, with a positive disrelish for his or her occupation, will be certain to make a failure of it.

The man or woman who lacks tact for any special business, had better (to use a solecism) give it up before commencing. And he who has no taste for the pursuit of poultry breeding, or "knack" in comprehending or managing machinery of any kind, should never purchase an Incubator, with the expectation of making a grand success with it. We meet such persons occasionally. They make—from descriptions or drawings—a machine, (or, perhaps, buy one), start it, and stock it with eggs, imagining that they have done all that is necessary; they await in delightful anticipation the expiration of the three weeks. It comes: possibly a few weakly chicks make their appearance, but more often it is a total failure. If blessed with more than the usual amount of patience and perseverance, another trial is made, with a like result; and the whole thing is given up in disgust, Artificial Incubation pronounced a failure, and Incubators a fraud.

We have in mind several instances like the above, one of which will bear reciting.

A fancier in a Western State had purchased an Incubator. Soon after receiving it, he wrote a very long letter, asking innumerable questions; some about matters with which it was to be supposed every one who had ever kept poultry was thoroughly conversant; others about the care of the Incubator, and which were already answered in the fullest possible manner in the directions accompanying the machine. The opinion was at once formed that that man had not the necessary "tact" to succeed with the Incubator. In less than a month, another letter to the seller of the machine requested him to either take it back or find a customer for it. "It would not work, could not be made to work, and was an outrageous fraud on the breeders". A few weeks afterwards, another letter stated that the machine had been sold at half price to a neighbor, who was having splendid success. Purchaser number one acknowledged he had been too hasty, had not had enough patience (a rare acknowledgement from purchasers of Incubators), was sorry he had sold it, and concluded the fault was entirely in his not understanding how to manage the machine.

The truth was acknowledged in this case, he had neither the knack or the patience necessary for success.

This "knack" consists not merely in understanding the mechanical apparatus, and the directions for the care of the machine, but in fully appreciating the necessity of attending to every point necessary to success and leaving no one of them undone. It is not enough that a machine shall work with almost a perfect equality of heat for the entire three weeks: there are other things needed, and the neglect of any one of them, small as it may apparently seem, may cost the loss of the entire clutch of eggs.

One of the first requisites is to have a room of comparatively even temperature, in which to run the machine. I say "comparatively", for I hold that an Incubator which requires a room of uniform temperature to ensure its successful operation, is but little better than a pan of water with a lamp under one end and a tray of eggs under the other. By comparative, I mean a temperature with a variation of not over fifteen to twenty degrees during twenty-four hours. Of course, no Incubator will work uniformly in a varying temperature of twenty-five or thirty degrees, because the heat necessary to keep up the temperature in a very cold room will make it too hot when a room gets very warm: and *vice versa*, the heat which is amply sufficient to keep eggs at the required degree in a warm room, will not nearly do it in a very cold one. We all know that a stove or furnace, used to heat any given space, must be regulated according to the amount of heat needed; and that, with an out door temperature of zero, it takes double the amount of coal that is consumed in moderate weather, to keep the room or space comfortable. Without attention and an increase

of fire heat in extremely cold weather, the room cannot be kept
warm. If run with the same draft and same supply of coal the
season through, the room will be too hot in moderate weather and
not warm enough in very cold weather.

The same philosophy that relates to the stove, applies with
equal force to the Incubator. No matter how much of a "self-
regulating" machine it may be, the supply of heat must be regulated by hand, giving more or less as the temperature of the room
requires.

The use and benefit of a regulating attachment is in controlling
the temperature of the machine, from day to day during the temporary absences of the operator. And that one which will do its
work under the greatest changes in temperature, and is the most
reliable under all circumstances, and yet simple enough to be
comprehended by every one, is the one that will be adopted by
the majority of those who use an Incubator.

No matter how simple, or how complicated the apparatus, it
must not be constantly fussed with. Do not put in any eggs until
the working of the machine is perfectly understood, and when
once comprehended, then expect it to take care of itself, without
frequent alterations.

Regularity in the necessary attention to the machine is important. The lamps had best be filled and trimmed daily, and that at
early evening, or even before it is dark, if it can be done conven-
iently. My own practice, when at home, is to fill and trim just before dark. Then leave the machine an hour or more; returning,
examine the lamp to see that the flame is of the proper height;
then take out the drawers, etc.; regarding which, see "care of
eggs". The directions must be carefully conned, and implicitly followed. Water trays or pans cleaned and refilled weekly, or oftener,
if necessary. Tank and boiler kept full; all soot, that may accu-
mulate, must be cleaned off daily. A very thin coating of soot
will destroy half the heating power of the lamp, which is equivalent to a consumption of twice the quantity of oil.

Where electricity is used, the battery must be watched, and
chemicals added when necessary; do not wait until the battery
has weakened, but keep up its strength. The zincs must be
cleaned as directed, and worn out ones replaced. It is best to
clean the zincs and jars alternately; i. e., not removing both cells at
same time; thus the battery is kept more uniform in strength.

An Incubator rarely gets thoroughly heated through, and all its
parts in uniform working order, under two or three days; hence
it is better, as a rule, not to put eggs into the machine until the
third or fourth day.

All the mechanism should be kept free from dust and dirt: heavy or gummy oils should not be used, they collect dust and make the machinery work hard. Oil only where directions say, and that with the best sewing machine or sperm oil. It is better not to oil at all, than to use thick or gummy oils.

Should there happen a very cold snap of weather, and the means of heating the Incubator not be adequate to keep up the temperature, cover it with a heavy blanket—an old army blanket will do—leaving only the ventilator and lamp uncovered. Should the heat have gone down, draw off the water and refill with very hot water. This will be found to be often necessary with some kinds of Incubators.

Padding, or hanging pads or cushions of feathers, felt or cotton, around the machine has been resorted to by some parties, and, it is claimed, with much success. The machine is stated to burn much less oil, to run more evenly and to be less subject to outside changes of temperature.

For heating purposes, only a high grade oil should be used. Kerosene of 150° fire test will be found the most economical and is perfectly safe. It is also much more free from smoke and from the strong kerosene odor so much disliked by many persons. I cannot give any special brand of oil, for it is sold all over the U. S. under different names. Ask for kerosene oil of 150° fire test. The ordinary kerosene, such as is sold by country grocers, is about 110° to 115°. A cheaper oil of about 90° is frequently substituted. These oils are the ones that so frequently cause the terrible explosions and burning accidents of which we daily read. Oils of 150° will not explode, and may be used with every assurance of safety.

When the season is over and the Incubator is to be laid up, the water pans, tank and boiler should be emptied, draining all the water off that will run. Then burn the lamp with a low flame for several hours, leaving all the stoppers open, so as to dry out the inside of the machine. After this, put the Incubator away in the attic or some dry room, until wanted again.

A frequent cause of trouble is using a higher flame than necessary. This causes the heat in the egg chamber to continue to rise —sometimes one or more degrees—after the ventilator opens. Only heat enough should be used to keep up the necessary temperature; all above that is wasted. It is much better to run the Incubator so that the opening and closing of the ventilator does not occur more than two or three times daily, rather than every half hour. Where these frequent changes take place, there is a waste of heat, and consequently of oil.

This waste of heat is peculiar to machines without regulators—

cheap machines, which depend on a very free circulation of air through constantly open ventilators, as a prevention against over-heating. The constant passing off of the heat necessitates a rapid and steady generation of the heated air, which must add largely to the consumption of oil.

With hot air machines, *i. e.*, those which use the heat direct from the lamp or stove, employing no intervening medium to re-tain and distribute the heat, there must either be imperfect venti-lation or a great waste of heat. If the ventilation is free, it re-quires a large consumption of fuel to keep up the required temper-ature. If the heat and fuel is economized, it must be done at the expense of ventilation, and if the latter, the drawer soon becomes foul and full of bad odors. At one of our large shows during the winter of 1880, the pleasure of viewing a very successful exhibition of hatching in a large glass machine, was almost counteracted by the vile odors which came from the badly ventilated apparatus. Very many visitors turned away in disgust, or viewed the ex-hibition beyond the smelling limit.

Very few, probably none of the many Incubators yet construct-ed, have an absolutely uniform heat in and over all parts of the egg drawer. The front part, that nearest the door, is usually the coldest; in some machines there being several degrees difference between front and back. Yet it is here, in front, that the ther-mometers are usually placed. It should be at the point where the heat is highest, as it is there where all the danger is. Some machines have a thermometer placed in the tank, giving the temperature of the water: this is never reliable, as a dozen causes may make the heat in the drawer vary greatly from that of the water. Others have the thermometer made stationary—fastened to the case of the machine: unless provided with a very long mer-cury tube, it cannot give the actual heat; it is too near the out-side of the machine.

In placing the thermometer, it should be where the heat strikes the egg in the natural process. Select the warmest portion of the drawer, and place the bowl of the thermometer as nearly as possi-ble on a level with the top of the eggs. If it is laid on the eggs be sure to have the stem or scale a little the highest—a very little will do, but it should never be perfectly level or with the bowl the highest; if so, the weight of the mercury more than counteracts the cohesion of the metal, which is all there is to draw it into the bulb as the temperature decreases, and the thermometer will most likely show a higher temperature than actually exists, and also, will not show any decrease in the heat of the drawer, if such should occur.

CHAPTER IV.

THE EGG.

Deeming that it is necessary to a full understanding of the process of Artificial Incubation, I shall give as briefly as possible, an account of the formation of the egg, and the mode and probable time of impregnation.

In the structure of the egg we have first the shell, which is sufficiently porous to allow the process of respiration. Next is the shell-membrane, which is double, and composed of an outer thick

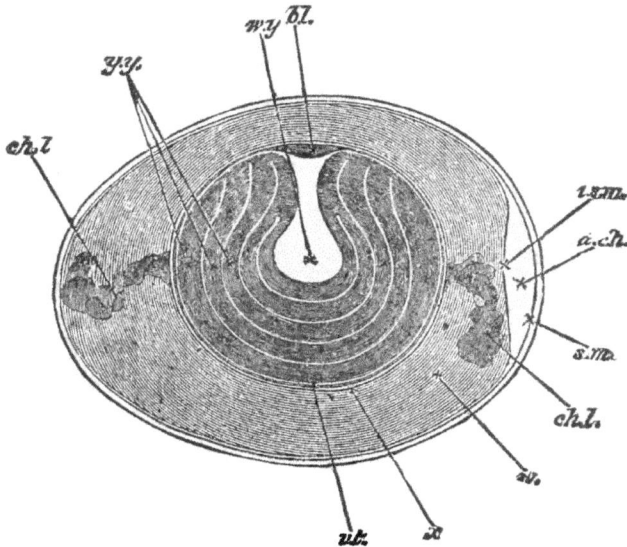

FIG. 1.

leathery membrane, (s. m. Fig. 1) and an inside thinner one (i. s. m.) At the large, or "butt" end of the egg, between the shell and the membrane, is the air-cell or air bubble (a. c. h.). In freshly laid eggs this space is so thin as to be hardly discernible, it rapidly increases in size, however, as the white of the egg shrinks from evaporation.

Next under the shell-membrane is the white of the egg or "albumen," as it is called (w.). This is formed of numerous thin layers of fluid albumen, which alternate in consistency, these form a network of elastic character, completely surrounding and protecting the yellow yolk (y. y.), which comes next, and in turn nearly encircles the white yolk (w. y.) The yellow yolk is also form_ed in several concentric layers, separated by very thin fluid lay_ers of albumen. This white yolk has the appearance of a round white spot of three-sixteenth to one-quarter inch diameter, connected by a funnel-shaped opening with the upper and outer edge of the yellow yolk, on the surface of which floats a small white disc about one-quarter inch in diameter; this is technically called the blastoderm or cicatricula (bl. Fig. 1). The layer of albumen immediately surrounding the yellow yolk, forms, at each extremity a spirally twisted cord (chl.), the outer ends of which are free and do not quite reach the shell. The action of these is not to suspend the yolk as many think, but rather to keep it in position by their weight and elasticity. They are called the Chalazæ. The entire yellow yolk is closely encased in a thin elastic sac, called the vitelline membrane (v. t.), which at the top doubles downward, passing around and under the white yolk. Examining the blastoderm with the naked eye, we can distinguish an opaque white rim surrounding a partially transparent centre, in the middle of which we see a small irregular shaped white spot.

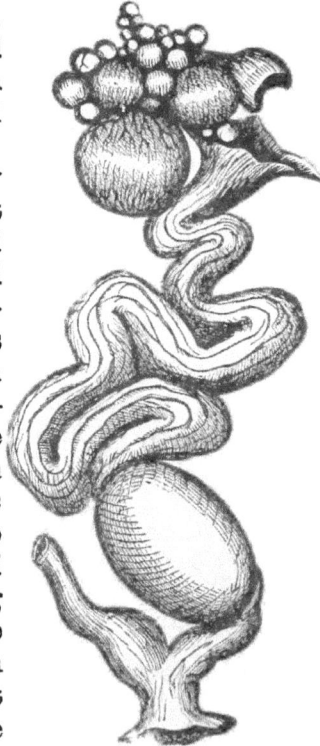

FIG. 2—THE OVIDUCT.

This is the spot at which impregnation takes place. This disc is known as the germinal disc, and the white spot as the germinal spot. In an unimpregnated egg the disc is marked with numerous irregular clear spots.

Before going further, it is necessary to speak of the formation of the egg, during its passage through the oviduct or egg-passage. (See Fig. 2.)

This, in a hen of ordinary size, is nearly two feet in length. It is placed directly under the ovary—a portion of which is shown in the cut—and as the yolks mature they break loose from their enclosing membrane, called the *ovisac*, and are received into the oviduct.

This latter, consists of four parts. 1st. The dilated proximal extremity of the ovary. 2d. A long tubular canal opening by a narrow neck into the 3d, which may be called the uterus; and 4th, the passage which leads from the uterus into the cloaca. It is in the second portion that the yolk receives its covering of albumen and the chalazæ. In the narrow neck above spoken of it receives its shell-membrane, and in the third portion the shell is formed. The passage through the second portion takes from three to four hours, and that through the third portion from twelve to eighteen hours. Impregnation occurs in the upper (first or second) portion of the oviduct. The spermatozoa is found here, moving in a fluid which the passage contains. It is not quite certain that impregnation can occur after the deposition of the albumen, though the fact that spermatozoa have been discovered in the albumen, makes it possible that they can bore their way through and reach the germinal disc. The blastoderm, as we have seen, covers the germinal spot like a watch-glass; as incubation progresses it spreads like a thin circular sheet over the yolk, immediately under the vitelline membrane, covering more and more of the yolk until it complete-

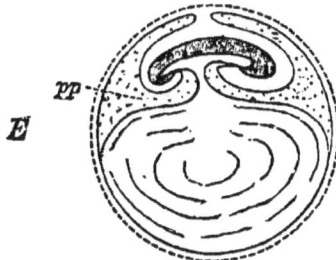

FIG. 3.

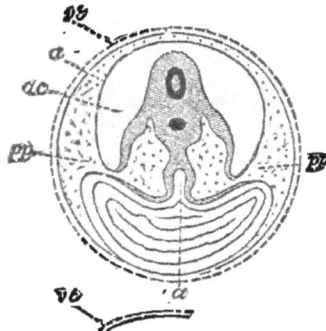

FIG. 4.

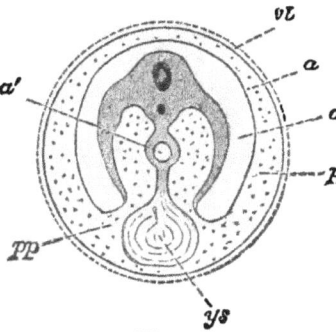

FIG. 5.

ly encloses it. In Fig. 3 we have a longitudinal section of the yolk with the embryo already formed. This may be seen with a powerful light, though not so perfect as it here appears, at the end of the first day.

Fig. 4 shows a section at thirty-six hours, and fig. 5 at forty-eight hours. In the latter the yolk sac (*ys.*) of the inner yolk is nearly absorbed into the alimentary canal (a^1) and the embryo begins to show quite distinctly.

Fig. 6 gives another view of the yolk (looking down) at the end of the third day. This shows the heart (H.) in the centre of the yolk, and the arteries radiating from it. By this time the yolk is about one-half covered by the rapidly forming veins and arteries, and the embryo may be seen through the shell of the egg with a good egg-tester. If a portion of the shell is carefully removed, directly over the embryo, the faint pulsations of the heart may be distinctly seen with the naked eye. By the end of the fifth day the yolk is completely enclosed by the blastoderm. The embryo appears in a curved shape; the feet making their appearance, looking like faint streaks on the end of the leg-like appendages. By the end of the sixth day, the heart, which was at first a simple tube, attains almost its complete form, and its covering—the pericardium—is first seen.

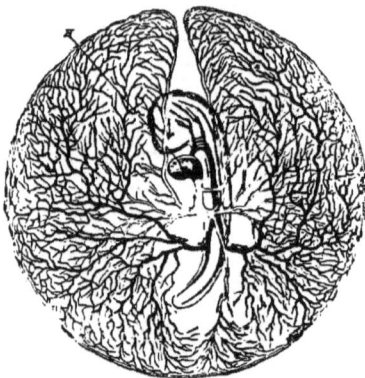

FIG. 6.

On the seventh day the crop and rudimentary intestines make their appearance, the beak begins to develop, and if the shell is broken, the first movement of the limbs is seen. A rocking or pulsating movement of the whole embryo may be seen through the shell, if held before a strong light. The yolk now becomes more fluid owing to the rapid absorption of the white. From this to the tenth day the embryo grows very rapidly; the bones beginning to form and flesh appearing on them. The embryo, at the eleventh day, if taken from the shell, appears as shown at Fig. 7. On the thirteenth day the toe-nails appear and also the formation of the scales on the toes; the feather-sacs have sufficiently developed to show the color of the coming chick. The fifteenth day a change of position takes place, the chick lying lengthwise in the shell; previous to this, the embryo has lain as formed—crosswise; the bill opens and closes, and distinct motions of the wings and legs may be seen. By the close of the sixteenth day the white of the egg has entirely disappeared. The yolk-sack connected to the chick by the umbilical cord lies loose between the body of the chick and the

FIG. 7.

shell-membrane, until the nineteenth or twentieth day, when it is drawn into the abdomen of the chick.

Fig. 8 shows the appearance of the chick on the nineteenth day with the yolk-sack not fully absorbed. And Fig. 9, the chick as he

FIG. 8.

lies folded in the shell just previous to exclusion. About this stage the chick pierces the membrane with his bill and begins to breath the air contained in the air-cell. The development from this time is very rapid. The blood ceases to flow through the outer membranes, which shrivel and dry up, the egg-sack is entirely absorbed, and the umbilial opening closed. The chick begins pounding at the shell of his prison, and soon makes for himself a way out. In doing this he turns in the shell; at first breaking a small hole, and from that cutting around the shell (see Fig. 10) until it yields to a strong push, and the loosened end flies off: resting

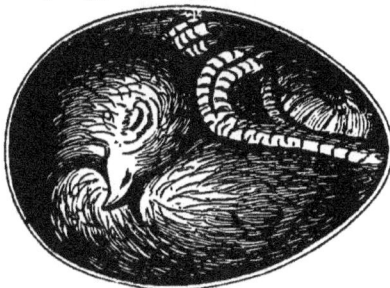

FIG. 9.

a few moments, he gives one or two more vigorous kicks, which clears him from the shell, after leaving it in the condition shown at Fig. 11, and he begins his active life. Here we will leave him for the present.

The fertility of the egg depends on a number of circumstances. The number of hens allowed to each cock, the size and vigor of the male, the breed of fowls, and sometimes on the individual bird.

With Leghorns, Hamburghs, Houdans, Games, and other small-bodied birds, from twelve to twenty hens to one cock will usually ensure fertile eggs; while with Cochins, Brahmas, Dorkings, etc., eight to ten should be the limit, and if kept in confinement, six will be found a safer number. With Crevecœurs, my experience is that the males are deficient in virility. In the Asiatic classes the size of the male may cause non-fertility. An unusually large bird is clumsy, and so heavy that his weight bears down the hens, and defeats, or prevents impregnation. In this connection may also be mentioned a frequent cause of trouble experienced by poultry breeders, and which,' for want of a better name, I have termed "Ineffective Fertilization." The cause of

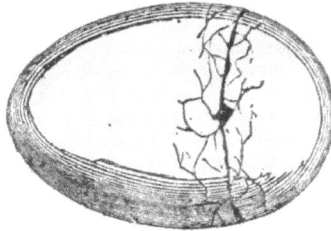

FIG. 10.

it I have been unable to determine. We find it first shown in eggs which have been under incubating heat for twenty-four to forty-eight hours; the germinal spot receives the impregnating spermatozoa, the veins form, pulsations begin and cease, and life dies out. Other eggs in the same nest, or side by side in the Incubator, progress, and in due time hatch out strong and healthy chickens. This ceasing of life not only occurs during the first twenty-four or forty-eight hours, but may happen on any of the twenty days previous to the exclusion of the chick from the shell. We find it at the fifth day, at the tenth, the twelfth, fifteenth, while other eggs under the same treatment in all respects, progress and hatch. It may be, and has

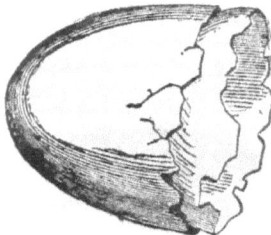

FIG. 11.

been charged, in the case of artificial hatching, to lack of moisture; too much moisture; too high a heat; not enough heat; bad ventilation; and half a dozen other assumed causes; but if it occurs only in a few cases. and these not in any certain part of the nest. or egg-drawer, but scattered indiscriminately among the rest—if these few cases are caused by an improper amount of heat or moisture, why not more? or all? If a whole clutch is gone, then the trouble probably is with the incubation. But these scattering cases, occurring as they do under the most favorable circumstances for the hatching of every fertile egg, are seemingly chargeable to no other cause than "Ineffective Fertilization."

Another matter, which may properly be considered here, is the time and limit of impregnation. Some fifteen years ago, I was convinced by the result of a chance *mes-alliance*, that if hens of one breed were allowed to run with the cocks of another their value as breeding stock for pure bred fowls was destroyed. The question has again and again formed a subject of discussion in our poultry journals, and has led me to make a careful series of experiments, the results of which I think of sufficient interest to present here. They were enough to show me that my first position was untenable, and again convince me that my first conviction was wrong.

I give the experiments in the order they were made. First, the time of impregnation. I took a Brown Leghorn pullet, which had not yet laid, and penned her by herself. The first five eggs laid were placed in the Incubator, and, as I anticipated, proved not fertile. A cockerel, of same breed, was placed with her at night, after she had gone to roost. The next egg laid (the sixth) was non-fertile, as was also the seventh. But the eighth, ninth, and tenth were fertile, and all hatched. At this stage the cockerel was taken out, and a Light Brahma cockerel, with heavy leg feathering, was substituted. The next six eggs were put in the Incubator; the eleventh and twelfth hatched pure Brown Leghorns, the thirteenth showed a little leg feathering, and the fourteenth, fifteenth, and sixteenth were all well feathered on the legs. Again the cock was taken away, and the eggs placed daily in the Incubator. The seventeenth proved non-fertile, the next four were fertile, and showed the leg feathering heavily. Then three more, with gradually decreasing leg feathers; the twenty-fifth germinated, but died at the fifth day, and the twenty-sixth to thirtieth all proved non-fertile.

I again took the same pullet and put a Brown Leghorn cock with her. The first day she did not lay; the second and third she did. The fourth she missed again; the fifth, sixth, seventh, and eighth she laid. All these eggs were put in the machine, and all proved fertile, except the one laid on the second day. I did not carry them through to hatching. This would seem to indicate that if a hen was laying at the time the cock had access to her, the third egg laid subsequent to intercourse would be the offspring of such male.

I tried the same experiment, later, as to time of impregnation, with a Brahma pullet, and in two trials had a result of fertility, in the first case, of the second egg, which was laid on the fourth day, and next in the third egg, which was also laid on the fourth day.

The next experiments I shall mention, were as to the limit of impregnation.

I took a laying hen from the yard, and penned her by herself in a grass run. The sixth egg, and from that on until the twentieth, which was laid the twenty-second day after her separation from the male bird, were placed in the Incubator. The sixth to eleventh eggs proved fertile; all after that, barren. Two hens were next taken from the yard, and put in same pen, and the eggs saved after the seventh day. Of these, one of every day's yield proved fertile, and of the other, fertility ceased with the tenth egg after separation from the cock. Still another trial of two hens resulted in the sixth egg of one hen and all after that proving barren, and those of the other, up to the twelfth, proving fertile.

I then selected a Brown Leghorn, a Light Brahma, a Houdan, and a cross-bred hen; after three changes of the cross-bred, I got four different shades of color in the eggs. The Leghorn eggs proved fertile to the ninth egg; the Brahma to the fourth, the Houdan to the thirteenth, (the last two, however, dying at two and five days), and the cross-bred to the sixth.

A Leghorn cockerel was then put with these four hens. The second Leghorn egg—laid on the fourth day—was fertile, also the first Brahma egg, laid on the fifth day, the third Houdan egg laid on the third day, and the second cross-bred hen's egg, laid on the second day.

This would give an average of fertility extending seven to eight days after the separation of the hen from the cock; but subsequent experiments lead me to think that it is not safe to rely on more than five or six days, except with the non-setting varities— Leghorns, Houdans, etc.

I know of one well authenticated instance of a Houdan hen's eggs having been saved, and set for upwards of twenty days after sale and removal of the male bird, and the eggs proved fertile up to the twenty-fifth day.

The next experiment I give was virtually a continuation of the first. The same Leghorn pullet, after having been with a Brahma cockerel three days, was penned with a cockerel of her own kind. The influence of the Brahma showed plainly in the first four chicks, was not apparent with the sixth, but showing again slightly in the eighth. Six eggs after this, to the fourteenth, were set, the twelfth and thirteenth only hatching, and both were without any apparent cross of Brahma blood. A second experiment of the same kind with a Houdan hen, crossed with a Leghorn cockerel, and then put with a Houdan cock, showed distinct marks of the Leghorn comb to the sixth, and slight indication in seventh and eighth egg hatched after removal of the Leghorn cock. I conclude from this that the average influence of a *mes-alliance* between birds of different breeds may be considered as commencing with the third, and terminating with the seventh or eighth eggs.

A NON-FREEZING WATER FOUNT.

CHAPTER V.

CARE OF THE EGGS.

Properly, this should commence with the laying of the eggs, and I will start from as near that point as possible.

Many beginners in the poultry fancy seem to think that eggs intended for sitting need a particular kind of treatment and manipulation. While there is no doubt that careful handling and protection frequently greatly add to a successful result, still it by no means follows that eggs will not hatch if subjected to rough treatment and exposure to cold.

So far as the latter is concerned, I would much rather the eggs be kept in a room where the temperature ranged from freezing point to forty degrees Fah., than in a place where the heat reached as high a point as eighty degrees. Of the two extremes, I would also prefer to trust the eggs to the lower; by which I mean, that I am convinced that any ordinary degree of cold—say ten degrees below the freezing point—does not necessarily kill the vitality of the egg, so long as the shell is not broken by the action of the the frost; and, on the contrary, the vitality of the egg is greatly injured and decreased by exposure to a temperature of eighty degrees, or thereabouts, for any considerable time.

Of course, I do not wish to be understood as recommending either of these two extremes, but would choose a temperature of about forty to fifty degrees as being the most suitable.

There seems to be no doubt but sudden, violent jarring will sever the chalazoa in the interior of the egg, and thus destroy its vitality; though eggs will stand the trembling, shaking motion of a wagon or car, and give the most satisfactory results in hatching.

The real care necessary is very little; simply gather the eggs daily, mark them plainly—variety and date—(washing them if much soiled), and lay them away in a cool room. If intended for use within two or three days, place them in a basket or other receptacle. If it is necessary to keep them longer, it is best to have flat boxes af about three or four inches in depth, in which place an inch or two of bran or sawdust—I prefer bran,—place the eggs in this, large end down, as close as you can pack them. An egg cabinet is a very convenient arrangement for storing eggs. This may be made cheap and simple, or as elaborate as one's fancy and pocket may dictate. Our illustration (Fig. 12) gives a very plain representation how one may be constructed. Some persons insist that the small end down is proper and preferable. My experience does not coincide with this method.

The theory is simple and easily understood. The air cell in the large end of the egg always enlarges while the egg is laid away. If the butt end is up, the evaporation of moisture is greater, and the pressure of air through the open pores at that end rapidly increases the size of the bubble. If the butt end is put down, the weight of the liquid contents of the egg, pressing down on the bubble, hinder the entrance of the air and allow only a very slow and gradual increase in the size of the air cell. This can easily be tested by taking two eggs of equal size, and carefully marking the rim of the air cell in each with ink: put them away, one point down, the other butt down. At the end of two weeks compare them. Eggs shipped for hatching should always be packed this way. For sitting, use only average-sized, regular-shaped eggs; un-usually large and uneven-shaped ones are apt to be

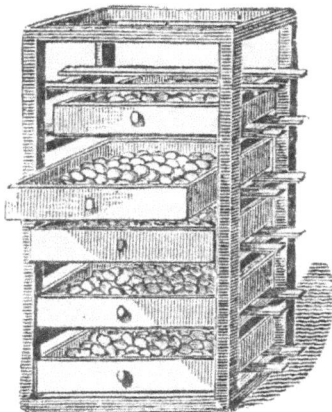

FIG. 12.

infertile, or to produce malformed chickens.

The amount of cold and exposure eggs will bear without injury during the term of incubation has created much diversity of opinion.

From a comparison of records of experiments conducted by the writer, and also from other data which I have compiled, I find that the ratio of loss decreases very fast as the process of incubation proceeds. Thus, when the eggs have been subjected to the necessary heat for forty-eight hours, the danger from chilling or overheating is vastly greater than at or after the fifteenth day. Eggs that would be ruined by an exposure of four or six hours to cold, or two or three hours temperature of 110° on the fourth day, will remain uninjured through double that time and greater heat or cold at the seventeenth day.

It frequently happens that a hen leaves her nest during the

early stages of incubation, and the eggs are perfectly cold when discovered. The hen may sit with exemplary steadiness during the rest of the time, but no eggs hatch. The same result may, and probably will, follow at any time during the first ten days, while during the last week an exposure of a whole day would, perhaps, seriously affect only a small number of eggs.

The same with an excess of heat; 108° is dangerous, and 110° almost sure death to life in the egg at any time during the first week if continued for half an hour or more, while I have subjected eggs to a heat of 115° on the seventeenth, and again to 120° on the eighteenth day, and yet hatched seventy-five per cent. of the eggs thus exposed.

From this data I have deduced the following ratio of liability of loss during the twenty-one days of incubation. By exposure to cold, we mean such as would ensue from the hen quitting her nest during April or May. Intense cold or heat, of course, not being considered, as that is always fatal to life: Eggs under heat for twenty-four to forty-eight hours will usually be affected the same as if perfectly fresh; at four days, the loss is about ninety five per cent;

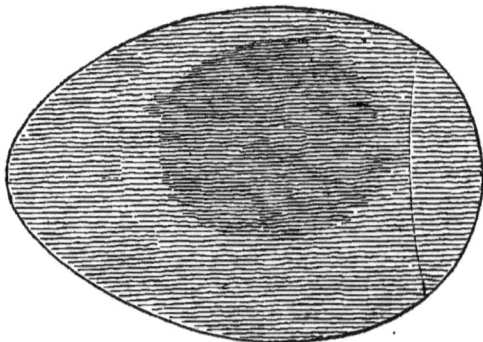

FIG. 13.

at six days, ninety per cent.; eight days, eighty-five per cent.; ten days, eighty per cent.; twelve days, seventy-five per cent.; fourteen days, sixty per cent.; sixteen days, fifty per cent.; eighteen days, twenty-five per cent.; twenty days, ten per cent.

Of course this cannot be accepted (nor is intended) as a positive table, applicable under all circumstances. The fertility of eggs varies with different breeds and under different circumstances, and while some are destroyed by very slight exposure, others will stand an extreme degree of heat or cold.

We now come to the egg under process of incubation. During the first three days it is best to turn them once daily; twice will do no hurt, but I am not certain that it is of any benefit. I have left them unturned for the first three or four days, and had a hatch of over seventy-five per cent. Sprinkling with luke-warm water once daily is sufficient for the first ten days; and for the first five days, I believe. if the air in the egg chamber is well charged with moisture, sprinkling will not be necessary. It is after the absorption of the albumen that sprinkling, or some other form of application of moisture to the egg, becomes positively necessary.

After the third or fourth day, the eggs must be turned twice daily to obtain the most satisfactory results. About this period— the third to fifth day—the eggs should be tested, and the barren ones taken out and their places filled with fresh ones.

The accompanying illustration (Fig. 13) shows the appearance of a barren egg, as viewed through the egg-tester. Such eggs do not change their appearance, save the enlargement of the air bubble, if left in the machine during the whole three weeks of incubation. After the eighth day the white, or albumen, becomes watery, and the yolk loses its

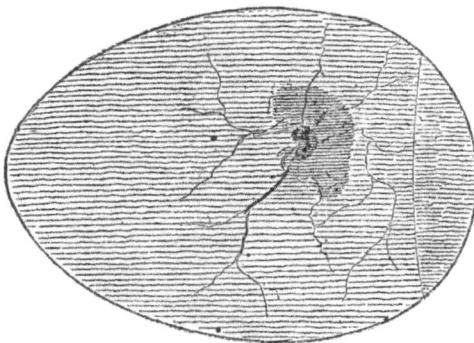

FIG. 14.

consistency; but a non-fertile egg will not become rotten by the simple action of the ordinary incubating heat. If removed at two or three days, it is not injured for culinary purposes; but if left under heat a week or ten days the white and yolk will usually run together if the egg be boiled. My practice is to save the barren eggs, taking them out at the first testing, and boil them for the young chicks when they are first hatched.

Figure 14 shows a live embryo at four to five days, as seen through the tester ; and Figure 15 a dead one, i.e., one that has germinated, but from ineffective fertilization, or other causes, has died. In these cases the dead germ, or

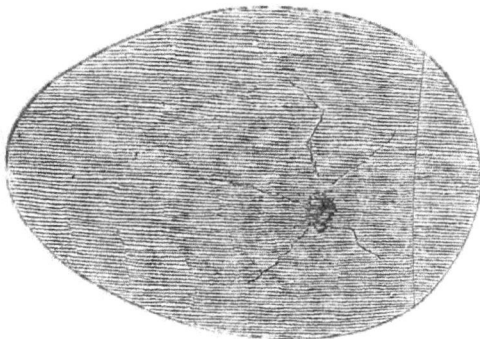

FIG. 15.

dark spot, will often be found adhering to the side of the shell, while in a live embryo the germinal disc will gradually rise to the top as the shell is turned. In the live embryo the veins will be distinct, and, if the egg is white-shelled, the color of the blood can be seen; in a dead one, the veins are simply dark, irregular lines, broken and clouded, the cloudiness shading into the vitelline membrane.

Those eggs which have germinated but died, should also be removed. These are the ones that become rotten, and are so offensive.

This testing should be repeated about the tenth day, when the living embryo will be seen through the shell, as represented at Fig. 16. Holding the egg steadily the chick may be seen to move in the shell—not suddenly, but more of a waving, or slow pulsa-

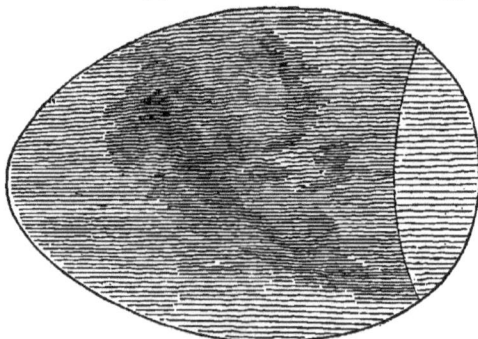

FIG. 16.

ting move-ment. The appearance of the dead embryo at this stage is shown at Fig. 17, without distinctness or form.

The final test should be made at about the fifteenth day, at which time those that were left in at the previous trial as doubtful will be found to have under-gone no change, and should be taken out. It is best to mark such eggs as we are in doubt about, with an interrogation point (?). They are then readily distinguished from those known to be fertile and alive. By this time most of the embryo chicks will be feather-ed, and will almost fill the shell, leaving a small, light space on one side of the small end ·of the shell, as also the air cell at the large end.

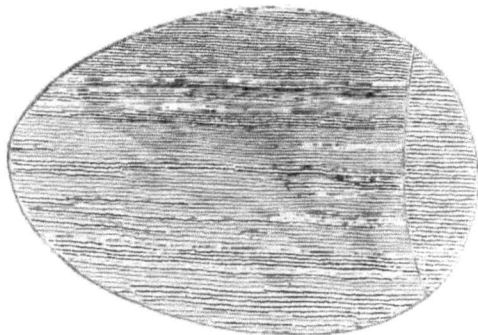

FIG. 17.

By the eighteenth day the entire shell is opaque, except the air cell at the buttend, and the egg appears as shown at Fig. 18. If, however, we find an egg appearing like Fig. 19, with the air cell on the side, and all else opaque. in nine cases out of ten the chick will be found to be dead.

We also often find an egg at about the eighteenth day with the air bubble occupying nearly one-third of the egg—the line between the chick and the air cell being irregular but distinct. In such cases the chick is either dead, or, from some cause, not as fully matured as it should be. If hatched at all, it is usually deformed; but generally it dies in the shell.

In these illustrations, prepared expressly for this work, I have given the general appearance of the egg as seen through the egg tester. Variations are frequent, the air bubble sometimes appearing to one side of the butt end; but I have never yet found a fertile egg with the air bubble, at the first testing, in the position shown in Fig. 19.

In turning the eggs it is well to change their positions in the drawer. Usually the center of the egg chamber is the warmest; in some machines one end of the drawer is several degrees warmer than the other. In nearly all, the front, nearest the door is the coolest. If the eggs are allowed to remain in the same position through-

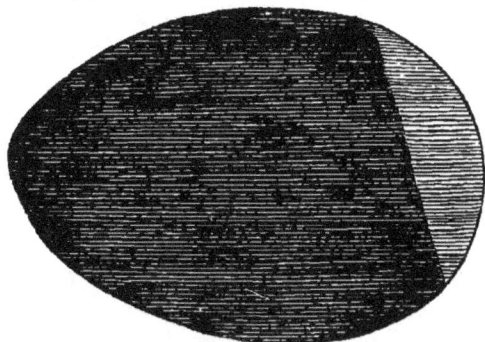

FIG. 18.

out the whole period of incubation some get too much heat and others too little. I am speaking now of the general run of Incubators. There are several makes that vary so little in the temperature in the egg drawer, that these remarks do not necessarily apply to them, although the practice recommended, can do no harm even with these. In changing their positions take those from the center of the drawer and roll those next into the vacant places, placing those taken out on the outside edges of the drawer. In this we are simply following nature's teaching. The hen when sitting imparts a greater heat to the eggs directly under the breast than

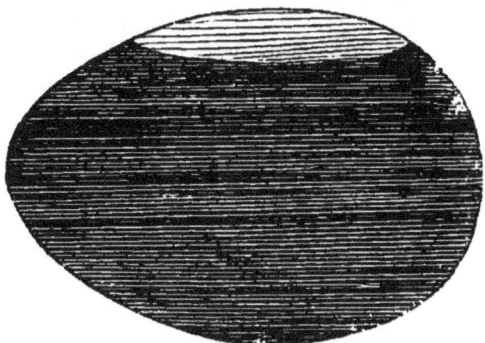

FIG. 19.

to those on the outside edges of the nest under her wings. Nature, or instinct, teaches her to move the eggs frequently, bringing those from the outside into the center of the nest. If any of my readers will watch a sitting hen from some point where the hen cannot see him, he will find this movement made at intervals of every half hour or oftener.

Mr. Wren, an English experimenter, tested this matter by marking the eggs. He says:

"I have examined the eggs under a sitting hen, and made the following memoranda of the position of the eggs during a part of one day. At 10.30 A. M., marked four eggs, and left them in the center of the nest; at 1.30 P. M. three of the marked eggs were moved to the outside. Marked three more, and left them in the middle; at 2.45 P. M. the three marked last were moved to the outside. Marked four more and left them in the middle; at 4 P. M. the four marked last were on the outside, and some of those marked first were back again to the middle of the nest."

The well known habits of the ostrich corroborates this fact. Her nest is made in a hollow in the ground, and the eggs, to the number of eighteen to twenty-four, deposited in it. Both male and female ostriches incubate, taking turns at the work. They sit astride of the eggs, their legs on either side of the nest, and the outer eggs are never changed in their places. The consequence is that about from four to six eggs in the middle of the nest hatch, and the remainder are addled. It was this large percentage of loss that led to the use of Incubators on the ostrich farms of South Africa.

From the tenth day onward, the eggs should be sprinkled twice daily. This may be done by the fingers, with a small whisk-broom, with a florist's rubber sprinkler, or by an "atomizer." I use about one gill of water to a drawer of one hundred eggs. It is best to apply the water when the eggs are first taken out of the Incubator, leave them ten or fifteen minutes so that they may absorb the water, after which turn them and replace them in the machine.

If turned, then sprinkled, and put at once into the Incubator, the heat of the egg chamber evaporates the water before the eggs can absorb it.

It is during the first ten days that the greatest percentage of loss usually occurs. A very variable heat, extremes of heat or cold, or any prolonged term of either, will kill the embryo at this time, when the same exposure after the fifteenth day would probably have no bad result. A variation of three and four degrees is not necessarily hurtful. If the air is moist, the eggs will bear an even greater variation of temperature; but if a dry air, it is generally injurious if allowed to get above 105° or 106°.

It will be found that a heat of 101° to 102° is the safest for the first week, gradually increasing to 104° at the close. If the eggs are being placed in the machine from time to time, a steady heat of 102¼° to 103° will be found productive of the best results.

Dead eggs will sometimes be overlooked and left in the machine; frequently, these may be detected by small drops of yellowish fluid, or of frothy liquid exuding through the shell; remove them as soon as discovered, handling them very carefully. Do not shake them or unpleasant results may follow.

About the eighteenth day I give the eggs a quick dip in warm water —say about 95° to 100° Fah. This is repeated on the nineteenth, twentieth, and twenty-first days, if the chicks are not coming out quick and strong. A good, healthy lot of chicks will commence pipping the shell on the twentieth day, and sometimes on the nineteenth, and by the commencement of the twenty-first, two-thirds of the clutch will be out. If the hatch is delayed beyond the twenty-first day they are apt to be weakly.

When the egg first shows the crack, or break, made by the imprisoned chick, it is a good plan to place it with butt end the highest; this gives the chick a better chance to turn, and to get out after cutting the shell all around. If the chick seems to be long in breaking out, the shell membrane being dry and tough, he may be helped. The best way to do this is to hold the egg with the hand immersed in a pan or bowl of tepid water, leaving only the bill of the chick, or the broken part of the shell, above the surface; then with the other hand carefully break the shell and pull off the enclosing membrane. This can, with a little practice, be successfuly accomplished in nearly every case.

It is sometimes desirable to know if the chicks are alive in the shell, and the probability of their hatching. During the final stages, the egg tester does not always decide this point. Two other tests are sometimes used—the water test and the glass test. The first may be used with perfect safety on eggs that have been under hens, but it is risky to try it on eggs that have been in the Incubator, the pores of the shell being so open that it takes but a short time to absorb water enough to drown the chicks. The test is to place the eggs in a pan of lukewarm water: those which sink are unmistakeably dead, those that float on the side are rotten, and those that have healthy, living chicks in them will soon bob around in the water in a very amusing manner.

The glass test, while not so thorough as the water, is always safe. It consists in placing the eggs on a pane of glass, lain flat a table or stand. The live ones will usually show a trembling, vibrating motion, or perhaps roll along the glass. The disadvantage of this test is that one cannot with certainty tell the dead ones. I have dropped one or two drops of cold water on an egg that had lain motionless on the glass for some seconds—which usually brings a protest from the enclosed chick as the chill strikes him—but without effect, and on breaking the shell found the chick alive, apparently all right.

A not uncommon complaint from users of Incubators, is the dying of the embryo at about the eighteenth day. The cause of this is hard to determine. It may result from a lack of moisture; though when this is the cause, death will not usually ensue until the twentieth or twenty-first day. In high altitudes, or any elevated and very dry atmosphere, this is a frequent occurrence and can only be obviated by a liberal sprinkling after the tenth day.

It is almost impossible to prescribe the exact amount of moisture necessary to give the best results, because the amount varies with the state of the outside atmosphere. It varies also with the temperature; more moisture being requisite in warm weather than in cold. Experience is perhaps the surest teacher in such cases, and even experience is sometimes at fault.

We sometimes have directly opposite results, under apparently the same conditions and management. For instance, A and B, each buy and use an Incubator of same pattern. The machines work to perfection during the whole three weeks: On the twenty-first day, A gets ninety per cent. of good healthy chicks, while B gets fifteen or twenty per cent., and sixty or seventy per cent. dead in the shell, at the eighteenth or twentieth day. Both have "followed the directions for management to the letter," yet with entirely different results. What caused B's failure? Who can say positively where his management was at fault. I have heard of two such cases the present season. It might be ascribed to lack of sufficient moisture; to too much moisture; too much heat; not enough heat, or a half-dozen other causes. If the directions have been implicitly followed—and we have no means of knowing that, for one person's interpretation of them may be, and frequently is, totally different from anothers—then the failure can be ascribed to only one of two causes: either lack of vitality, or unfavorable electrical condition of the atmosphere.

I am inclined to think that lack of vitality, or defective fertilization, was in this case the cause of failure; otherwise, why did not the remaining seventy per cent. hatch as well as the twenty? If seventy per cent. died in the shell from any mistake in management, why not all? Clearly, because some were more perfectly fertilized than others, and having more vitality, lived through the unfavorable stages and came out.

CHAPTER VI.

After the chicks are hatched, they may, in some machines, be left several hours; or even a whole day, in the egg chamber. With others it is necessary to remove them as soon as the feathers are dry. The directions that go with the Incubator must be the guide in this matter.

Either brooders or hens must be in readiness to take care of the chicks. If brooders, there are three or four different makes from which to make a selection. For indoor use, where there is no trouble from rats, any of them will answer a very good purpose; for out-door use we recommend such as is represented by Fig. 20 (description of which will be found in a succeeding chapter.)

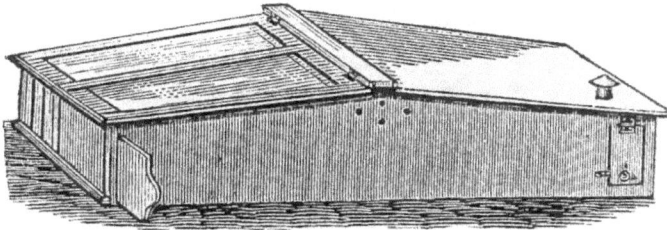

FIG. 20.

If hens are to be used to brood the chicks they should have been kept sitting some days in advance; when the chicks are taken from the Incubator, which should be at evening, they should be put under her and allowed to remain all night. In the morning take her off with the brood, and if she is quiet and gentle, more chicks may be given her from the machine. In very cold weather I should limit the number to fifteen; in May she may have thirty or forty and will take good care of them. Put her in a good, dry, clean coop. Figs. 21 and 22 show a plain and cheap style which most breeders can make themselves.

If anything more elaborate is desired, specimens are shown in Fig's 23, 24 and 25. The last two are shown with wire runs attached.

This is of great value where cats and rats show too great a liking for chickens, and may be used with great benefit until the chicks are ten days to two weeks old.

With the brooder it is especially valuable, where they are placed out doors, in keeping the young chicks near home until they have learned where to run for shelter and warmth. The runs I have in use are of different shape from that shown in the cut. They are four feet long, two wide, and one high, made of three-quarter inch mesh No. 18 wire netting, fastened to a frame of band iron, all galvanized. Something like the one shown at Fig. 25.

Another excellent run may be made of wire netting two feet wide, cut into six foot lengths and tacked

FIG. 21.

to wooden frames, two by six feet; an end piece two feet square will be necessary if it is wished to cover the run. The same frames may be made of quarter-inch round iron, the ends on one side projecting four to six inches, and pointed so as to stick into the ground and hold the frame in an upright position. Where the wooden frames are used it will be necessary to tie them, or hold them in place with hooks.

For the first twenty-four hours no food need be given; the yolk-sac that was absorbed by the chickens just previous to their exclusion from the shell, affording all the sustenance re-

FIG. 22.

quired. For the next forty-eight hours give them the yolk of hard-boiled eggs, crumbled or chopped fine; stale bread, moistened in milk is also good. Oat-meal, though rather too expensive for frequent use, is one of the best kinds of food. It should be soaked in water or milk before using. During this time they should be fed little and often—as often as every two hours. Water may be given them in dishes that should be so protected that the chicks cannot get into it. I say "may be given," for it is by no means necessary. If wet food is used I would as soon refrain from giving water until they are ten days or two weeks old. I have raised as good broods without water as with it.

After the third day a variety of food may be used. That of which I use most is coarse Indian meal and wheat bran, equal parts by measure, thoroughly mixed and scalded, to which I add, if I can get it, about one-eighth as much ground beef scraps. If the latter is not to be had, I would throw in a gill of fine ground bone to every fifty chicks can be had, give week a feed of sheep's, or swine's fine. With the which the liver the meal and bran it as soon as it "dry scald," not This meat food is necessary early in the chicks can get

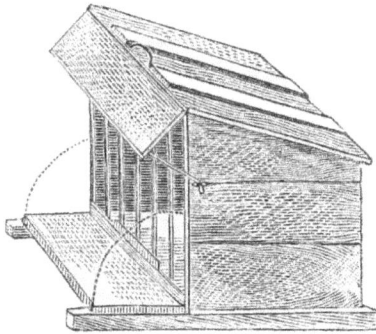

FIG. 23.

When neither them twice a boiled beef's, liver, chopped hot broth in was boiled, scald mixture, and feed cools. Make it a mushy or pasty. more particularly the season before earth-worms and

snails. They need animal food of some kind, and in the absence of worms, bugs, and insects, the ground scrap-cake is a most excellent substitute.

In giving mixed, or wet food, it is advisable to use feeding vessels of some kind. The one represented at Fig. 26 is a good contrivance; sizes may with the rying in different kens. The and outer may be wood or feed cups made in like cut-

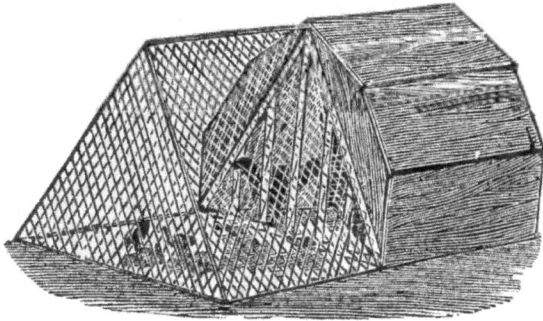

FIG. 24.

several be made wires va- width for aged chic- bottom rim of top of either metal; the may be sections, ting a pie

into quarters. Our illustration represents alternate ones for feed and water. They are filled through the top, which lifts off.

For water, when it is given to the young chicks, we prefer the simple apparatus shown at Fig. 27. It is home-made, and can be made by anyone with a pair of tinner's snips and a couple of old preserve cans. Cut off the bottom of one an inch and a half high; the other, as shown in engraving, cut three inches up, making the openings three-quarters of an inch wide for the first chicks. After a week or ten days others will have to be made one inch wide; bend these ends in, or cut them off, and put together as shown in the cut. Another style of water vessel is shown at Fig. 28, which is also of tin or iron, with round holes one inch diameter cut in before putting it together. A conical-shaped cover fits on top. Still another is shown at Fig. 29: an inverted preserve can in a flower saucer, with a few old nails or bits of stone un-der the edge, or two or three Λ shaped notches around the edge.

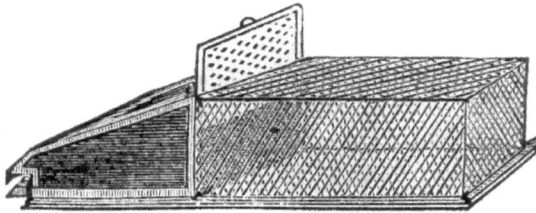

FIG. 25.

Fig's 30 and 31 are stone-ware drinking fountains, very highly recommended by many who have tried them. All these watering arrangements should be placed in the shade, or protected in some way from the hot rays of the sun; and they should be cleaned out and refilled regularly, morning and night.

To return to the food:—one very highly recommended is "hoe-cake"—Indian meal stirred into water, and then put into the oven and baked. Crumble it up, and feed it the same as grain. Chickens thrive most excellently upon it. The French have a sys-tem for sup-plying ear-ly hatched chicks with animal food which in some sec-tions, might be used with profit. They make pits in the ground, about three feet deep, siding it up with bricks or boards

FIG. 26.

(an unused hot-bed would answer). Into this they put six inches of fresh horse manure, trampling it well down, and wetting it by pouring in pails of water; on this they place a layer of slaughter house refuse—blood, scraps of meat, intestines, etc.—two or three inches thick; then another layer of manure, and again a layer of refuse, until the pit is full; it is then covered by a foot of manure, and boards placed over that. In about three weeks time the pit is alive with maggots, which are fed to the chicks. Where much of this is desired, a succession of pits, ripening one after the other, must be made.

At two weeks old, begin feeding cracked corn, wheat or buckwheat at night. Up to this time the chicks should be fed six or seven times a day—every two or three hours; after four weeks old, three times a day will be sufficient.

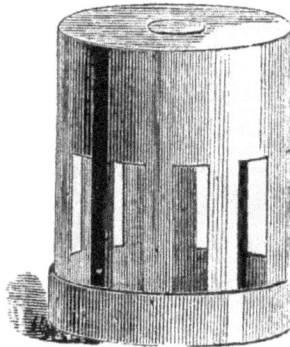

FIG. 27.

In addition to this "menu", there may be added, occasionally, feeds of boiled rice; small potatoes, boiled and mashed, and mixed with a little barley meal; bread crumbs and stale bread, steeped in milk or water; in fact, all the scraps from the table should be saved and given to them as soon as they are old enough. On the farm, where milk is abundant, scald the meal with the hot whey of curded milk, mixing the curds in with it. Into this a little red pepper and powdered charcoal may occasionally be put. Vegetable food should always be given at least every other day; lettuce, cabbage leaves chopped fine, a few carrots minced fine, onions also chopped, may be mixed with the scalded food, or given alone, until a week before killing, when the latter should be discontinued.

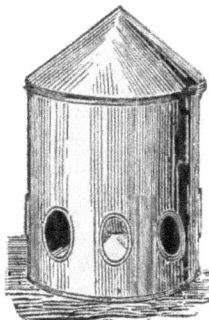

FIG. 28.

Lettuce may be raised in shallow boxes (called "flats" by gardeners), that can be started in any sunny window, and if placed at night out of reach of frost, can be grown without any artificial heat. Have plenty of fine gravel within their reach, and also some lime rubbish, old mortar, or broken oyster or clam shells. If near the seashore, get a few bushels of shells, throw them on a pile of brush and burn them; then break them up and put them in the yards. The value of table scraps, as food for both old and young fowls, is rarely fully estimated. If in the vicinity of a large hotel, the breeder would be well repaid even in buying them and carting them home.

The success of Mr. Warren Leland in his poultry raising was largely due to this supply of food. At that time he was one of the proprietors of the Metropolitan Hotel in New York City; all the scraps from which, amounting to from fifteen to twenty barrels weekly, were sent to his farm and fed to the poultry and hogs.

Dust baths are a necessity to the health of the chicks. Take shallow boxes, and put into them three or four inches of fine road dust or fine sand, in which mix a few handfuls of ashes, or the dust of tobacco leaves. Some recommend sulphur. It may be used, but very sparingly. Air-slaked lime is much safer, and nearly as effectual in killing or preventing vermin.

While chickens can be grown on a bare run, still, grass runs are much better, and, if possible, they should have a few hours on the sod every day. A good supply of green food of some sort, as advised above will in some measure supply the lack of grass, and in winter and early spring has to; but later it had best be provided, if one's facilities will admit it.

FIG. 29.

The bare ground, if many chicks are raised, will soon become contaminated. I advise turning it over frequently with a spade or plow. If with the spade, a certain portion daily, or half the plot semi-weekly. It is also a good practice to reserve a corner, and sow oats, buckwheat, or clover on it (fencing it off from the rest of the plot) and letting the chicks on it when a few inches high.

A very excellent prac-tice, which I once saw in use in Clinton, Mass., is to plant a portion of the yard with toma-toes, and turn the chicks into it just as the fruit is ripening. The vines furnish both shade and food which is highly, relished. If the yards, or runs, are not well provided with shade trees, I advise putting out low-grow-ing evergreens, currant and gooseberry bushes; peach and plum trees will also thrive well in the poultry yard, and bear yearly crops of fruit. A quick shade may be made by mak-ing a low, rough trel-

FIG. 30.

lis, and training quick-growing vines, such as morning-glorys, or gourds, over it. Rough, scraggy brush stuck into the ground will soon be covered by these vines, and make an excellent pro-tection from the hot summer sun.

Crowding must be avoided. This is a stumbling-block upon which many trip. It is not safe to put over seventy-five chicks into one brood (or one partition), and it is much better to limit the number to fifty. At three weeks old these should be reduced in number to thirty or thirty-five. It is safer to divide the brooder into two or three parts, and put twenty-five in each, than to have them all together. During the day they may all enjoy the same run, but at night they should be separated as above.

Be careful in the use of the brooder not to give too much heat. The body heat of the hen is only 98°, and if the brooder is so arranged that the chicks can get their backs against the warm tank, or pipes, they should not have more than that amount of warmth. If the chickens come out in the morning, looking as if they had passed through a sweat-bath, they have been kept too warm, and it will not take many repetitions of it to decimate the brood. The temperature of the brooder where the chickens nestle is much increased by the animal heat of the chicks themselves, and after the brood is a week or ten days old very little artificial heat is necessary, except in quite cold nights or days. After the first of May, the young chicks really need no heat beyond the first week. Even during April, unless it is an unusually cold and wet season, the chickens may be transferred at three weeks old to a summer brooder. Such a one as is illustrated and described at Fig. 32 will be found well suited for the

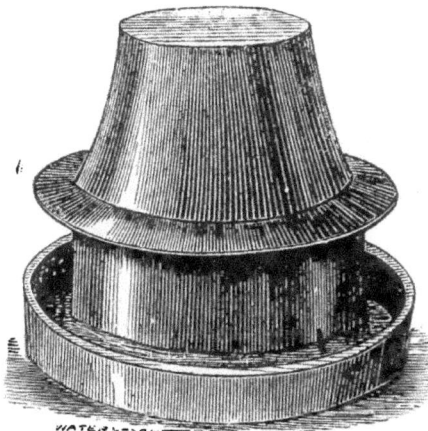

FIG. 31.

purpose; or the brooder previously mentioned may be used without the heating arrangement.

Cleanliness is imperative; the brooders must be cleaned often, and the bottoms covered with dry earth or sand. Dampness must not be tolerated in the quarters assigned to the young broods; it will be fatal to success. Feed cups and water vessels must be kept clean and sweet. Regularity in feeding must be observed; a glut one day and a famine the next will soon bring disease. Never over-feed; give what will be eaten up clean and no more. Have regular hours for feeding, and feed at such times. Vary the food to suit the weather; in cold, damp weather give more animal food than in warm, dry days; it is more nourishing. Do not allow any portion of the work to be slighted. Remember, the better you attend to them the better they will pay you. What is worth doing at all is worth doing well; and, finally, always keep in mind these three requisites to success: Warmth, Cleanliness, and Regularity in Feeding.

CHAPTER VII.

REARING CHICKENS TO ADULT AGE.

Thus far we have treated of the chickens as being raised to the age of broilers—six weeks to two months old.

When carried beyond that age, either with a view for future market or breeding stock, they should be given more liberty, more range of run. No matter how rough the land, provided it is not infested with minks, weasels, rats, and other poultry enemies: a newly cleared piece of woodland makes an excellent run.

For summer accommodations I would advise the erection of low sheds. These may be cheaply built, as follows: Set six posts into the ground, making a parallelogram six to eight feet wide and

FIG. 32.

twelve long; on the top of these posts, six feet from the ground, spike a plate piece of 3x5 inch timber, all around. Three feet below this, halve in another piece for the floor; roof it over with either tongued-and-grooved boards, or with rough pine boards laid like clapboards; let the peak of the roof be three feet above the plate; floor it with common pine or spruce boards; the sides and ends may be of lath, placed one inch apart, or of wire cloth, two-inch mesh. A door should be placed in either one or both ends, with an inclined ladder from the ground for the birds to get up. The roosts, or perches, should be on a level, two feet above the floor. Whitewash the entire building inside and out. At night the doors should be closed, and the ladders raised, and the chicks are then safe from all four-footed enemies. They may be left to roost in such a shelter until the frosty nights of early November in this latitude, and will be all the better for it.

To have them grow and thrive they must be continually employed. A handful of wheat or oats in a sheaf of straw scattered over the ground will do them as much good as two feeds of the best ground corn when thrown down on a hard and bare run. Chickens must never be allowed to mope or be idle; they should be kept continually exercised. Better let them be a little scantily fed so that they will search for their food than to stuff them into laziness. When nearly time to market them fattening foods and methods may be used with good effect. You have then a frame to build upon, and the food produces the wished for results in the shortest possible time.

Tincture of iron may now be given in the drinking water, or else use iron kettles for the drinking vessels; it helps the growth of bone and feathers.

In feeding, pursue about the same system as recommended in the preceding chapter, giving more whole grain though as the birds get older.

The *menu* cannot be too much changed. Boiled vegetables with the meal mixed up dryly makes excellent feeding stuff; bread crusts and broken biscuits soaked in skim milk make chickens grow wonderfully fast.

Green food is more important than most breeders acknowledge. Unless they have unlimited supplies of grass or vegetable food I should feed it to them often.

Lettuce leaves and lawn mowings, finely minced up and mixed with some soft food, with a sprinkling of some tonic food, will often induce chickens to eat when nothing else seems to tempt them. Spiced foods and aromatic compounds I do not at all advise for young stock. I think they do more harm than good, however useful they may be for stimulating in adult birds. Chickens so fed seldom make fine specimens, but become matured when they ought to be growing, and develop combs when the birds ought to be making flesh and bone.

Water, too, is a very important item to their well-being. Chickens must have clean water, and they must have fresh water. I do not mean that stale water would kill a chicken, but I firmly believe that water which stands from morning to morning does them harm. Rain water which is allowed to stand in the water tub does them harm. Water which can be found by the chickens in stagnant puddles, or which drains from farmyards, does them harm, and when you give either food or drink which is hurtful they cannot thrive as they should.

In fattening and preparing for market a very different treatment must be given.

To give them delicacy of flesh make their principal food, for a week or ten days before killing, barley meal moistened with milk, occasionally alternating with Indian meal scalded with either water or milk. During this process the chicks had best be kept confined in a darkened room. If the business is conducted on a scale sufficiently large to warrant the expense, I should advise the use of a fattening house, with apparatus for "cramming". This process is better adapted to stock say six to eight months old, but may be used with profit on broilers from the sixth to the eighth week.

I have in use a stack of fattening pens which, built on a larger scale, will be found very convenient. They are constructed under a shed, against the back side, are in three tiers, ten in each tier; the apartments are each six inches wide, thirteen high and fourteen deep. The floor of each projects four inches in front, forming a shelf on which to place the food and water cups. An open space, four inches wide, is left at the back, through which the droppings pass to a trough on the ground underneath, which should be supplied with dry earth to absorb the moisture. Under the back part of the bottom tier is placed an inclined shelf, which throws the droppings forward into the trough. The front is formed of wire rods, three to each pen, passing through holes bored in top and bottom of pens.

Such pens ensure quietness, which is one of the essentials to quick fattening. The pens are large enough for a fowl to stand up or sit down, but not roomy enough for them to turn around or exercise. There is no fighting or restless exercise, and the fowls gain more in one week in such quarters than in two as ordinarily penned for fattening.

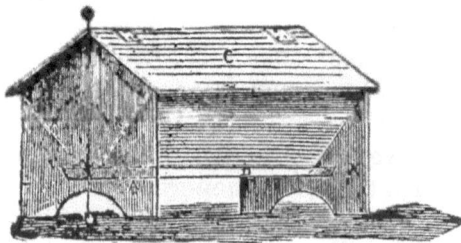

CHAPTER VIII.

ESTIMATES OF COSTS AND PROFITS.

In considering this occupation as a business, it will be necessary to vary somewhat from our text, and introduce matters which, at first sight, may seem foreign to our subject; but without these means of comparison we could not so plainly illustrate the benefit and profits of the artificial method.

The only practical illustration of Artificial Incubation as a business venture, which we have in this country, is that of Mr. Baker, of New Jersey.

His establishment is without doubt the largest in the world. It was fitted up and completed more like the hobby of a rich merchant than the ideas of a practical business man. Everything was built and finished expensively; woodwork all planed; galvanized wire netting dividing the runs; houses and runs fitted up like a gentleman's poultry yard for breeding exhibition fowls, and for showing to his wealthy visitors. The hatching house was most completely equipped with every convenience—heated by steam or hot water, electrical communication with all the egg drawers and the hatching room; and all seemingly regardless of economy in construction. The entire place, consisting of hatching house, twenty-five by fifty feet long, two stories and basement; nursery—a double-pitch glass house, one hundred and fifty feet long, and thirty feet wide, fitted with stationary brooders and runs, each six by twelve feet; the second house for more advanced chickens, four hundred and fifty feet long and twenty-four wide, with wire enclosed and covered runs outside; fattening and cramming house, fitted with revolving fattening pens, holding about two thousand fowls; slaughtering house; packing house; ice house, and roosting sheds, all said to cost in the neighborhood of eighty thousand dollars.

When I visited the place, the capacity of the Incubators was fifty-six hundred eggs. Allowing a result of four thousand chicks from each hatching, and twelve and a half per cent. loss in rearing, it gives a total of thirty-five hundred chicks every three weeks, or fifty-nine thousand five hundred per year; these at an estimated average price of forty cents each will give a gross return of twenty-three thousand, eight hundred dollars.

Against this, we have interest on investment, at six per cent. forty-eight hundred dollars; taxes and repairs, two per cent. more, sixteen hundred dollars; wages, six men, at five hundred dollars each per year, three thousand dollars; feed, five thousand dollars; coal, ice, etc., six hundred dollars, making a total of fifteen thousand dollars, and leaving a profit of eight thousand, eight hundred dollars for the year's operations.

I believe that the building and fixtures necessary to raise and market that number of chickens may be constructed and fitted up at less than one-fourth, and probably one-fifth, the amount said to have been expended above. Some of the conveniences and labor-saving arrangements might have to be dispensed with, as the extra labor involved would not probably equal the interest of investment.

I give the above as illustrating what may be done, even with the disadvantage of carrying such a heavy load of interest. With a more moderate outlay, the profits would be proportionately larger.

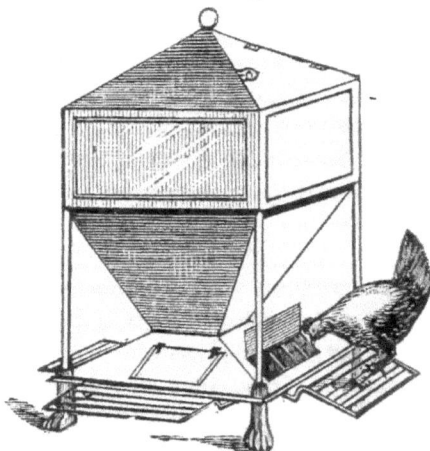

SELF-ACTING FEEDING HOPPER.

The average cost of hatching and raising a chicken by artificial means to eight weeks old, is not over fifteen cents. At this age it should weigh (if forced) from one and one-half to two pounds.

The cost of the next two months will be fully fifteen cents more, while the weight will not gain over one pound. It follows, therefore, that the most of the profit comes from the first two months' care and feeding. Broilers at that age, in New York, Philadelphia, or Boston, will sell during April, May, and June, at from seventy-five cents to one dollar per pair, and sometimes as high as one dollar and fifty cents per pair early in the spring. At this rate there is a large profit on the early-hatched broods; and prices for good broilers (plump chickens under two pounds each) will usually hold up to fifty and sixty cents per pair until October.

During the late fall and winter months, an Eastern breeder cannot hope to compete with the Western producer. Adult fowls can be raised there and marketed here (during cold weather) at much less figures than we at the East can do it. It is only by improving the quality of the birds marketed, and getting them into market at an early age, that we can make a paying business of it.

Bear in mind always that the profit on chickens is made by getting them fit for market at the earliest possible age.

To do this, the breeder must avail himself of artificial means. It is an absolute necessity to success that he employ Incubators and brooders in connection with the natural method. I say "in connection with," for I do not advise the use of artificial means to the entire exclusion of the natural. .Taking, for instance, a business in which it is proposed to keep five hundred hens. These, at an average of one hundred and fifty eggs each, will give seventy-five thousand per year. Now, allow a hen to sit twice, thirteen eggs being then used thirteen thousand eggs, and have sixty-two thousand which we are obliged to otherwise dispose of. With six Incubators, of two hundred eggs capacity each, we could use all of this surplus and produce a total of fifty-three thousand, five hundred and fifty chickens. Allowing each hen to hatch ten chicks per sitting—twenty chicks in all—we have ten thousand, making a total of sixty-three thousand, five hundred and fifty.

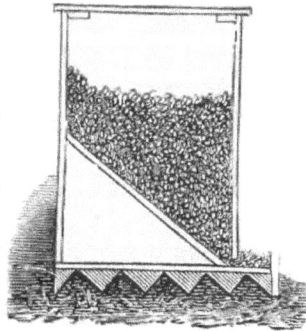

FEEDING HOPPER.

Now, the care of the six Incubators need not take over two hours per day, while the proper care of the hens, the necessary watching to see that they do not get off their nests, feeding, watering, cleaning the apartments and nests, etc., would take fully as much time, and, if only one hundred hens were sitting at once, the time would extend through seven months, with a return of less than one-fifth that of the Incubators. In this we have given the hens credit for about seventy-seven per cent., while the average hatch with the hens where large numbers are kept and set is not over sixty per cent. Figuring at this latter percentage, we should get only seven thousand, eight hundred chickens, against thirty-one thousand, two hundred and thirty-seven which would be the equivalent for the Incubators for the same length of time. The Incubator we have credited with a

hatch of eighty-seven and one-half per cent., while ninety and
ninety-five per cent. are now frequent averages with the Incubator that the writer is using.

Returning to the natural method, we have five hundred hens,
each of which is expected to hatch two broods of chickens, one
hundred of which are supposed to sit and hatch every three weeks.
Allow each hen to cover twenty chicks, we have fifty coops to provide every three weeks. Each hen will run with her brood four
weeks, and early in the spring the chicks must stay in the coop
until six weeks old. For winter raising, which must be done, a
house will have to be prepared, and the hens and coops kept in it.
This will necessitate the use of one hundred and fifty coops, which
at two dollars each will cost three hundred dollars. To this may
also be added the cost of the house. If portable brooders were
used, and the chicks taken from the hens as soon as hatched, and
put into them, it would require twenty-five (25) brooders, which
would cost three hundred and seventy-five dollars. These can be
be used in an open shed. The cost would be about equal in both
cases, while the labor in caring for them would be largely in favor
of the brooders, in that with the coops there would be one hundred to be fed and shut up every night, against twenty-five of the
brooders. Then again, with every coop and old hen, there is as
much dirt and filth to be cleaned out every few days as there will
be in a brooder for a month; the hen fouling the coop more in one
day than a brood of fifty chicks would in a week. In addition to
this gain, is that of the use of the hen for five or six weeks, during
which she may be made to produce twenty or twenty-five eggs.

But discarding the portable brooder, and using a house as sug-
gested above, the entire expense for the accommodation of the
chicks hatched by that number of hens need not exceed five hundred dollars. The saving in labor alone, over the hen and coop
system, would pay a large interest on the investment.

To go into it as extensively as the above-mentioned estimates
would necessitate, would preclude the use of hens and coops in
rearing , and the artificial method would have to be adopted; and
while the first expense would have to be comparatively large, it
would require a very small annual outlay to keep everything in
working order. The saving of chicks from casualties, from neglect, by the hens pecking and tramping, has been found to more
than repay the outlay, even in small poultry establishments. In
large enterprises, the percentage of loss is always more, and this
saving would consequently be greater.

CHAPTER IX.

An egg-tester is almost indispensable to the breeder who hatches by artificial means. It may be the simplest construction imaginable, or an expensive and complete apparatus.

Of the first, perhaps that described in *Wright's Illustrated Book of Poultry* is the most simple: "A plate of tin or zinc to shade the light from the eyes, with an aperture cut in it the shape of the egg. The egg is held to the aperture, with the light brought closely to the other side ".

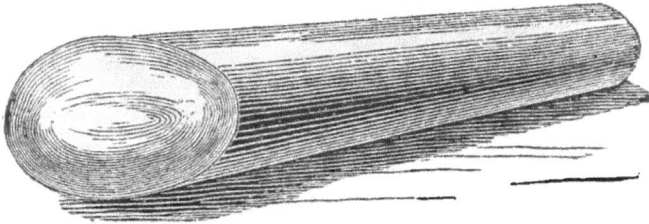

FIG. 33.

Next, perhaps, in simplicity is that shown by Fig. 33, which was devised by the writer about four years ago. It is made of a stiff piece of paper (dark color preferred), five inches long, six inches wide at one end and four and one-quarter at the other. This is rolled and joined together with a lap of half an inch, by either paste or a needle and thread, as shown in the cut. With this simple instrument the eggs can be examined at any time of day, and under any ordinary light. Its use is like a telescope—apply the smaller end to the eye, and hold the egg at or in the larger.

An improvement on this has lately appeared. It consists of putting a piece on the small end like the mouthpiece of a fireman's trumpet, and fastening a piece of some flexible material over the large end, with an egg-shaped hole in the middle. Instead of paper, the tube is made of tin.

Fig. 34 illustrates a simple arrangement which may be made out of a small cigar box or a stiff paper box of similar size. On the bottom of the box place a piece of looking-glass of nearly or quite that size; remove a strip from one end of the top, about an inch wide, and replace it at an angle of forty-five degrees, as shown in the engraving. Then cut one or more holes in the remaining portion of the top, about one and a half inches in diameter. The entire top (or else one end) should be loose, so as to get at the glass when it becomes soiled or dusty. To use it, put the eggs over the holes, and with the top of the box under a strong light, look into the opening, shielding the eyes by the strip set at right angles. The eggs will be reflected in the glass, and will appear clear or clouded, as the case may be.

This last article is more particularly adapted to household purposes for testing eggs for culinary use. Bad eggs may be immediately detected by its use, and rejected. For testing eggs under incubation, however, it is practically valueless, as barren eggs only can be positively told by its use.

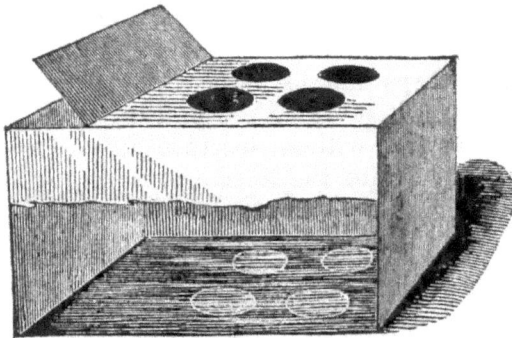

FIG. 34.

A more elaborate article is represented in "The Centennial Egg-Tester"—Fig. 35. This is a tin or metal case, nine inches high and six square, with a kerosene lamp inside, backed by a reflector; a tube of four or five inches long is opposite the reflector, on the end of which is a flexible covering with an egg-shaped hole, against which the egg is held. In a darkened room the interior of the egg is very distinctly shown, and the life of the embryo may be traced until the egg becomes quite opaque.

The "Utility Egg-Tester", Fig. 36, is another desirable style. An ordinary policeman's or bull's-eye lantern is the formation of the apparatus. A tube of several inches long is placed over the lens, and another is telescoped over that, the latter having the usual flexible cover and opening. By moving the outer tube in or out the light may be focussed on the egg, and its interior made very plain.

The outer slide being detachable, leaves the lantern free for use about the house or place.

Every fancier who has given it a thorough trial has arrived at the same conclusion as ourself, that there is no way of raising chickens so healthful and economically as by means of the Artificial Mother or Brooder.

This will, before many years, be regarded as an indispensable attachment to every poultry yard. With the fancier in a small way, who can keep only a few hens, it will be popular, because he can take the chicks from the nest and place them in the mother, and either re-set the hen, or turn her out, and in a week or so have her laying again. Another advantage is in the saving of coops. The young chicks may be put into the mother as they are hatched, not necessarily together, but at different times and of different ages—one, two or three weeks apart.

The large breeder also will find an advantage in this last respect, and an additional one in the item of feed. Young chicks need, and do better on a finer feed than the fowls. It may be desirable to feed hard boiled eggs for the first day or two, or some other equally expensive food. Where the hen is with the chickens she usually eats fully one-half of this specially prepared food. The breeder of Games will find it of the greatest possible use; in one season he would gain the value of several "mothers" in the saving of many of his young chicks from being picked to death by savage hens with broods of their own. How often many of us have felt like wringing the neck of some cross biddy upon finding her savagely picking (perhaps for the tenth time or more) the head off of some unfortunate chick, belonging to another brood, which had strayed into her coop!

FIG. 35.

Cleanliness is another very important consideration. One hen will soil the coop more in one day than fifty chicks in a week, and the droppings of the former always carries with it more or less waste of food.

Quick growth is desirable in all breeds, and nothing promotes it so much as warmth and freedom from vermin. The absence of the hen secures the freedom from lice, if the chicks are greased when taken from the nest, and if a warming attachment is used the chicks will huddle under the fleece during cool weather, remaining out only long enough to eat and drink.

We often see a hen with her brood of chicks standing around the coop, all drawn up with the cold, peeping and crying to be brooded, while the old biddy is intent only on filling her crop, or fretting at her confinement; and when allowed to range, many is the brood of chicks entirely lost by being dragged through the wet grass in the early morn, until one after another drops off, wet and chilled through.

Brooders, or Artificial Mothers, are not a recent invention. Reaumur, in 1777, used an "Artificial Hen" made of a box, one end of which was provided with a sheepskin fastened on an incline; the box was covered with horse manure, leaving only one end open for the entrance of the chickens; a small portion of the top next the open end was covered with glass, so as to give light and air to the chickens.

About the commencement of the present century M. Bonnemain, another Frenchman, devised an Artificial Mother, in which hot water was used as the heating medium; this was conveyed through the "Mother" in four pipes, placed a very little above the level of the floor. Flannels were attached to them, so fixed as to furnish a soft body for the chicks to get their backs against. A little later M. Carbonnier devised an aparatus, of which Figs. 37 and 38 are illustrations. Referring to the cut, A is a zinc tank or case for warm water; B, the tube for filling the tank, and C, the flannel

FIG. 36.

nel (in this case a piece of sheepskin) under which the chickens hover. The top of the case or box, in which the tank is placed, is of glass, arranged to slide, so as to open at pleasure. Three holes are provided on each side for ventilation, and a door at the end to keep them in when desired. This was the pioneer of the "*Hydro Mothers*". The trouble of refilling the tank several times daily with hot water was the principal objection to its use.

The Centennial Brooder, shown on page 37, Fig. 20, was invented by the writer about twelve years ago, but was not made in its present shape until a few years since. It consists of a tank and boiler, the connecting pipes so arranged that the water is kept in constant circulation. A small kerosene lamp heats the water in the boiler. The tank is placed in an inclined position, with a flannel cloth below it for the chicks to nestle under. The whole is enclosed in a wooden box, with a glass cover over the front, and a metal one over the tank and lamp. Ventilation is provided on all sides. The cost of using a Brooder of this style, accommodating one hundred to one hundred and twenty-five chickens, is about one-half cent per day for oil. It can be used in the open air, as the chicks are perfectly protected by the glass and metal covering, and the lamp so placed that the wind does not extinguish it. The tank does not need to be refilled more than once or twice during the season. It is manufactured by the writer at Rye, N. Y.

A somewhat similar affair is shown at Fig. 39—the Graves Artifiicial Mother. This was provided with a so-called self-regulating attachment, intended to prevent the heat rising too high. For a while this apparatus was fairly successful, but the regulator getting quicker out of order, it was soon laid aside. The "Perfect Brooder" is another late invention, for which the inventor claims the only

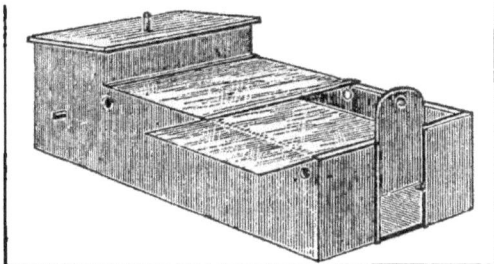

FIG. 37.

perfect imitation of the hen nestling her brood. It is a long, low box, with glass doors on top—a miniature hot-bed frame, in fact. Running lengthways, inside the box, are several iron pipes, through which a circulation of warm water is kept up by means of a small boiler at one end, heated by a kerosene lamp. Under the pipe is the usual brooding cloth. This apparatus is seemingly an adaptation of Bonnemain's Artificial Moth-

FIG. 38.

er.

The Eclipse Artificial Mother, manufactured at Waltham, Mass., is another invention for the same purpose, and is shown at Fig. 40.

The manufacturer says of it: "A full description of our Artificial Mother is unnecessary, as the cut above shows exactly what it is. They are even of more importance to a large breeder than an Incubator, for by using them a much larger per cent. of chicks can be raised.

"The essentials in an artificial mother or brooder are: First, a provision for furnishing the proper heat above the chickens; second, a good method of ventilation; third, a perfect freedom from

harboring-places for vermin, and a simple arrangement of attachment by which the fleece or woolen lining may be removed and
cleansed readily at any time.

"The Eclipse Artificial Mother meets all these requirements; and
it is the simplest, most portable, easiest managed machine of any,
for this use, that we have seen. The heat may be supplied or
withdrawn at any time, and its form is such that chickens brood
in it with perfect immunity from the stifling and crowding which
is the bane of many artificial mothers".

Among for-
ions in this
Reaumur's
ing appara-
ed by Fig. 41.
of a small
by an oil or

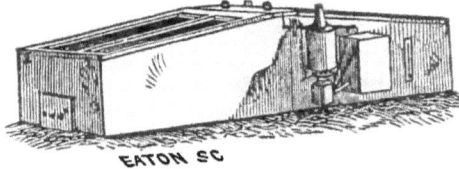

FIG. 39.

eign invent-
line we have
hot air brood-
tus, illustrat-
This consists
stove, heated
spirit lamp.

the heat being caused to radiate under a sheepskin before escaping; a circular rim covered with glass surrounded the heating attachment, and outside of this were placed long, movable boxes,
which could be covered with wire or left open on top. It answered
a very good purpose, but was too expensive for general use.

FIG. 40.

Mrs. Frank Cheshire's Artificial Mother is shown by Figs. 42,
43 and 44, the first being a front view, showing the curtain, which
is made of narrow pieces of woolen or flannel. S, Fig. 43, is the lamp
which burns naptha through a long tube inserted into an arched
chamber of the tank, the flame of which is shown at E, Fig. 44.
A B, Fig. 44, is the tank, filled with water, over which is a covering of wood, felt, or other substance. KK shows side views of
the brooding cloth, hanging in strips like the curtain in Fig. 42.
This brooder is very highly commended by English fanciers.

Christy's Hydro Rearing Mother, illustrated by Fig. 45, is the latest English invention in this line. In principle it is similar to the invention of M. Carbonnier, the heating part being a large metal tank, which is to be filled twice daily, or as often as necessary, with hot water. Under the tank the chickens nestle, protected on the outside by curtains of flannel cloth. An enclosed run surrounds the "mother" proper, of sufficient height to keep

FIG. 41.

the chicks within bounds. This is provided with sliding doors, to allow them exit. The tank is filled through the pipe C, and the cold water drawn off through the faucet at B.

These comprise about all the brooders worthy of special men-

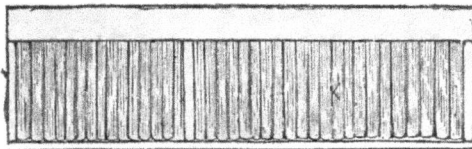

FIG. 42.

tion. There are a few other kinds, but so little different from those already described that they need not be illustrated.

The value and benefit of the Brooder to the fancier and breeder is aptly described by Mr. L. Wright, in these words:

"First, great economy of hens, as immediately after hatching they may have a second lot of eggs given to them, or be at once returned to the breeding house; in which case the first eggs laid

wlll be fertile, while, if a hen brings up her chicks, she passes three
weeks at least with them before laying, and her eggs are valueless
for hatching until four or five days after her return to the breeding pen. Second, economy of food, as all eggs, grits, and other
dainty food chicks. Special very young given in a through which
ones cannot pass.

goes to the dainties for ones are easily feeding coop, the larger

FIG. 43.

Hens, besides consuming much and destroy-
ing more, often prevent their broods from taking that which is
thought most desirable for them.   Third, economy of labor in
feeding and there is always commodation days. Fourth, ances, since their train of ain and feed-dispensed

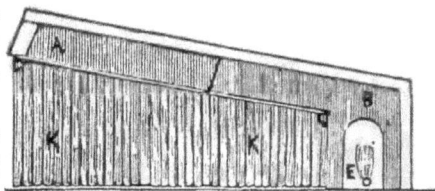

FIG. 44.

cleaning, while capital dry ac-for all on wet saving of appli-coops, with drinking fount-ing dishes, are with. Fifth,

the extreme tameness of chicks. A hen often prevents her brood
feeding till the attendant has gone; but, under this system, little
chirpers of three or four days old will run and flutter up to who-

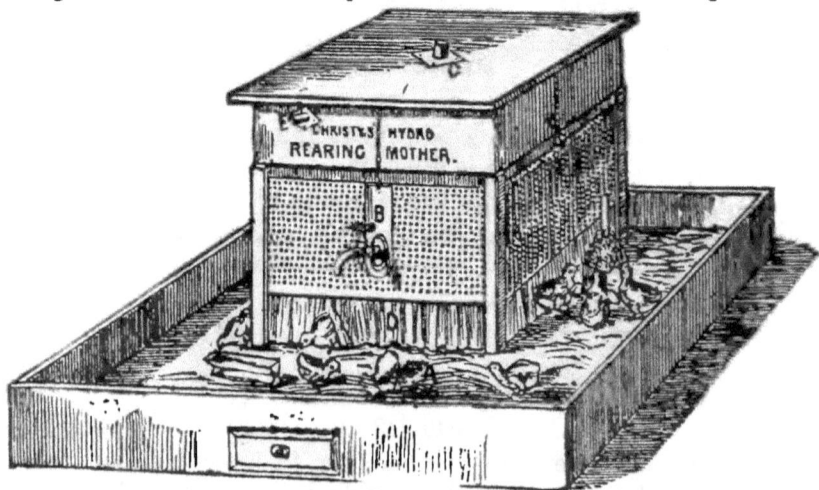

FIG. 45.

ever has the charge of them while they never seem to quarrel or
fight. Sixth, the accidents to which little nestlers are subject,
from the timidity or clumsiness of the mothers, are entirely
avoided. How many have to bewail chicks trampled to death;
while often sickness, terminating fatally, is but the result of some

internal injury, of which the hen has been the author. Seventh, increased size. Notes carefully taken of weights and ages during four seasons show that, during 1870 and 1871, by the natural system our chicks only attained the weight in twelve weeks which, during 1872 and 1873, under the artificial system, they reached in ten, though this partially is due, no doubt, to selection. Eighth, better feathering and stronger health, arising, probably, from nestling as often as desired. The second is proved by our loss of one chicken during 1872, and of not even one during 1873. Ninth, increased cleanliness of chicks, whose beautiful down retains its utmost purity until it is replaced by feathers. Tenth, the possibility of raising broods very early in the year, since they can nestle until eight or ten weeks old, if they will ".

CHAPTER X.

The earliest record we have of hatching by artificial means is mentioned by Herodotus, about 450 years B. C., in his reference to the egg ovens of Egypt.

They are also mentioned by subsequent historians, but it is not until A. D. 1494, nearly two thousand years after, that we find anything like an intelligent description of how they were constructed. In that year Alphonse II., King of Naples, established an Egyptian Incubator, and during the same year the Duke of Florence imported an Egyptian who was skilled in the art, and constructed an incubator after the Egyptian pattern. Neither of them were successful, however, owing to the difference in climate between the two countries. These

EGYPTIAN EGG OVENS,

according to the early descriptions, were built of mud, or adobe; in later years they made them of brick, which were, in fact, sun-dried mud. They are described as consisting of two parallel rows of small ovens, and cells for fire, divided by a narrow, vaulted passage; each oven being about nine or ten feet long, eight feet wide, and five or six feet high, and having above it a vaulted fire-cell of the same size, or rather less in height. Each oven communicates with the passage by an aperture large enough for a man to enter.

Fig. 46 gives a sectional view of the passage way and the ovens on each side. These are all under ground, and connected with the outer air by a long, ceiled passage, so as to avoid cold drafts. The small circular openings seen in the cut are about three feet in diameter; each orifice, or mouth, leads by a short arched passage into the oven. These ovens are not quite circular, but nearly so; the roofs are domed, and contrived with a kind of chamber over them; the apertures leading to the fire-chambers are the same width as the openings to the ovens, and only high enough to admit a boy to pass through. From each fire-chamber there is likewise a communication with the oven to which it belongs. In the domed roofs of the ovens, and in the roof of the room, there are holes that can be opened or closed at pleasure; these serve the twofold purpose of letting out the smoke, and letting in air and a dim, hazy light.

Fig. 47 shows the process of heating the ovens. The material employed for heating is called *gelleh*—dung collected and dried for the purpose, which is kept smouldering slowly in the fire-chambers above the eggs. Water is supplied in troughs made of mud bricks, encircling the eggs.

The climate of Egypt is specially suited for this method, in being almost of uniform temperature, and the men who follow the business are bred to it from childhood. This, as the *Maamal* of later days, demands constant attention, and the attendant on the ovens literally *lives* in them during the time they are in operation.

FIG. 46.

This egg-hatching is said to be carried on only during the months of April, May and June. The eggs are supplied by the peasantry, and there are two systems of purchase. Under one system, the hatcher pays down an agreed sum to the peasant for eggs; under the other, the owner of the eggs leaves them with the hatcher at his own risk, the latter agreeing to return one chicken for every two eggs.

According to statistics given during the last decade, the business is still one of large national importance, the number of establishments for the hatching of fowls' eggs in Lower Egypt being given as one hundred and five, and in Upper Egypt as ninety-nine. The number of eggs hatched in Lower Egypt is 13,069,733, and the number spoiled 6,255,867. In Upper Egypt the number hatched is 4,349,240, while the entire number spoiled is 2,529,660. In several works which refer to this subject these ovens are called "*Maamals*".

This is incorrect, the latter name being correctly applied to a peculiar portable stove sometimes used for the same purpose.

In China artificial hatching has been practiced for centuries, and probably as long or longer than in Egypt. The stories told of the system of incubation there seem hardly credible, and, although probably describing what the narrators saw and heard, are deficient in some points that were kept from their knowledge.

In "Minturn's Travels", he says: "On our return from the gardens we stopped at an egg-hatching establishment. This was a large wooden, barn-shaped building on the river bank. The eggs are purchased out of the produce boats that come down the river, and are here artificially hatched. The process employed is singular, as using only the natural heat of the egg, and is as follows: Large baskets, each twice the size of an ordinary barrel, and thickly lined with hay to prevent the loss of heat, are filled with the eggs, and then carefully closed with a closely-fitting cover of twisted straw. The eggs are now left for three days, after which they are removed from the basket and replaced in different order—

FIG. 47.

those eggs which were before on the surface being now on the lowest tier. At the end of three days more the position of the eggs is again altered, and so on for fifteen days, after which time the eggs are taken out of the basket and placed on a shelf in another apartment, and covered with bran. In the course of a day or two the chicken bursts its shell and makes its way out of the bran, being at once taken charge of by an attendant, who is always on the watch. The whole secret of the process is in the fact that the the animal heat of the whole mass of eggs being retained by the basket, which is formed of materials that do not conduct caloric, is sufficient to support the animal life of any one particular egg, and to foster its development".

CHINESE HATCHING BASKETS.

Another traveler describes what he saw as follows:

"The hatching house was built at the end of the cottage, and was a kind of long shed, with mud walls thickly thatched with

straw. Along the ends and down one side of the building were a
number of round straw baskets, well plastered with mud to prevent them from taking fire. In the bottom of each basket there
was a tile placed, or rather the tile forms the bottom of the basket.
Upon this the fire acts—a small fireplace being below each basket.
Upon the top of each basket there is a straw cover, which fits
closely, and which is kept shut while the process is going on. In
the centre of the shed are a number of large shelves placed one
above another, upon which the eggs are laid at a certain stage of
the process. When the eggs are brought they are put into the
baskets, the fire is lighted below them, and a uniform heat kept
up, ranging, as nearly as I could ascertain by some observations

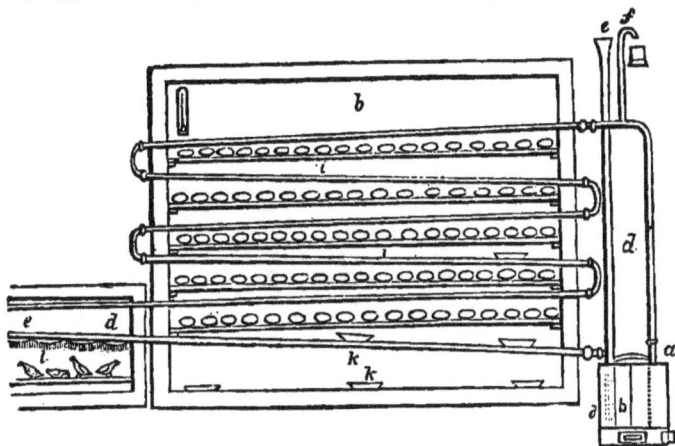

FIG. 48.—BONNEMAIN'S INCUBATOR.

which I made with the thermometer, from ninety-five to one hundred and two degrees—but the Chinamen regulate the heat by
their own feelings, and therefore it will, of course, vary consider-
ably.

In four or five days after the eggs have been subject to this
temperature they are taken carefully out to a door, in which a
number of holes have been bored nearly the size of the eggs; they
are then held one by one against these holes, and the Chinamen
look through them, and are able to tell whether they are good or
not. If good, they are taken back and replaced in their former
quarters; if bad, they are of course excluded. In nine or ten days
after this—that is, about fourteen days from the commencement—
the eggs are taken from the basket and spread out on the shelves.
Here no fire heat is applied, but they are covered over with cotton, and a kind of blanket, under which they remain about fourteen days more—when the young ducks burst their shells, and the
shed teems with life ".

There is but little doubt that part, and that perhaps the principal portion, of the hatching process was kept from the sight of
these narrators, for it is exceedingly improbable that any process
so directly antagonistic to the natural system could be successfully carried out. Doubtless there was some unseen means of
keeping the eggs warm, as it is impossible that the egg should of
itself possess sufficient heat to sustain life, much less develop it.

REAUMUR'S HATCHING APPARATUS.

No further record of any invention for artificial hatching comes
to our notice until the year 1777, when Reaumur, the celebrated
French naturalist, constructed his apparatus for hatching by
means of horse manure, except a "little portable oven" described
by Oliver de Serres, a noted French agriculturist, as being heated
by four lamps and the eggs covered with feathers. Of the date
of this, however, we have no knowledge.

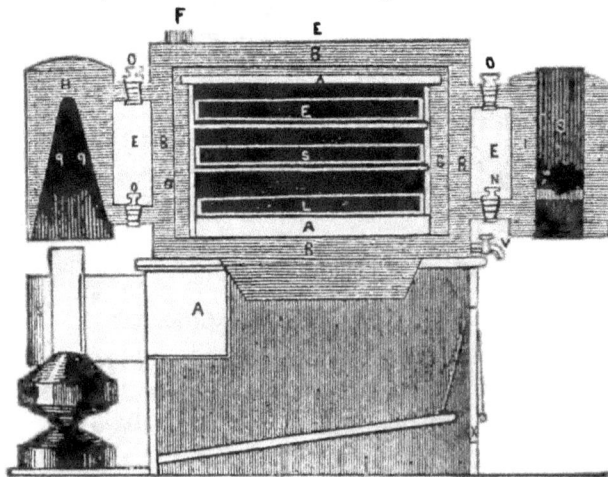

FIG. 49.—AMERICAN EGG HATCHING MACHINE.

Reaumur's apparatus was quite successful in the hands of the
naturalist, but with others it did not do as well, probably owing
to lack of attention and knowledge of the requisite care. His
apparatus was simply wooden casks fitted with drawers or movable shelves, on which the eggs were placed, the whole surrounded
with fresh horse manure, which was renewed at intervals to keep
up the heat.

BONNEMAIN'S INCUBATOR.

The next invention we find is that of M. Bonnemain, who was
the first to use hot water to warm the eggs. Fig. 48 is a sectional
view of his machine, in which a represents the boiler; b the box

or room in which the heating apparatus is placed; *d* the tubes for circulating the hot water; *e* the funnel end of the supply tube; and *f* an exhaust pipe to carry off the steam should the water get too hot: *c* is a box through which passes an extension of the lower coil of pipe, under the returning tube of which is fixed a piece of sheepskin for a brooder for the chicks. M. Bonnemain also used sponges saturated with water in the bottom of the egg chamber, to supply moisture to the air. He also constructed a regulating bar, formed of two different metals, (probably iron and brass), which acted upon a damper in the furnace door, thus increasing or decreasing the draft of the fire.

The eggs were lain on slides as shown in the illustration. The

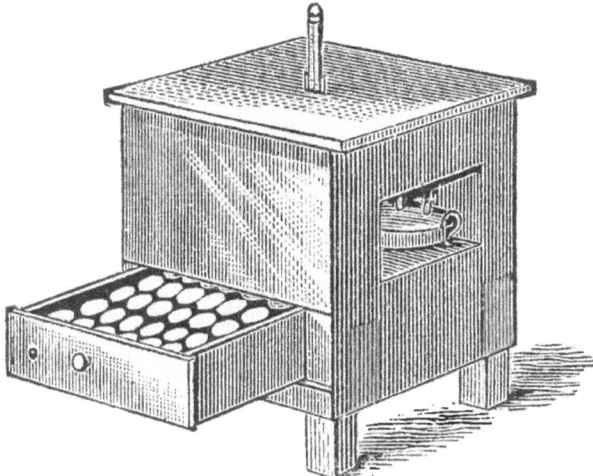

FIG. 50.—CARBONNIER'S INCUBATOR.

Incubator was not a success, owing to the impossibility of keeping the temperature even on the different slides.

THE ECCALEOBION.

Next after Bonnemain's, as near as we can tell, was an invention which was shown in London, and called an Eccaleobion: this was heated by steam pipes, with jugs of water in the egg chambers to keep the air moist.

In 1842, a small machine was exhibited in operation at Bristol, Eng., by a Mr. Appleyard; we have no description of it. About the same time one was exhibited by Mr. E. Bayer, in New York and Brooklyn, called the

POTOLOKIAN.

And between this and 1845, a Mr. L. G. Hoffman of Albany, invented and had in operation, the

AMERICAN EGG HATCHING MACHINE,

of which Fig. 49 is an illustration. This is a box two and a half feet long and two wide, enclosing a metal tank or cistern of water, *R. R. R.*, which is connected with the boiler on the left of the tank by two tubes. The boiler bottom is a long cone reaching nearly to the top, which gives a large heating surface to the spirit lamp which supplies the heat. The egg drawers are entirely enclosed and surrounded by the hot water cistern. The tank is filled through the tube *F*. The machine stands on a box, in which may be noticed an inclined board, the under side of this is lined with sheepskin and serves as an artificial mother. The end of the board is held up by a weighted cord, and may be raised or lowered to suit the size of the chickens.

CANTELO'S INCUBATOR.

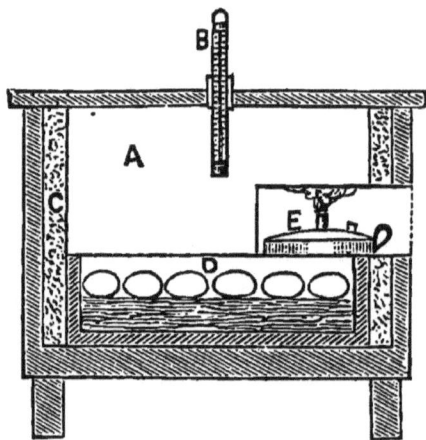

FIG. 51.—SECTIONAL VIEW CARBONIER'S INCUBATOR.

Cantelo in England, showed a fairly successful machine about this time. It was the first attempt to imitate the natural process of applying the heat from above. It is described as a very simple apparatus: being a tank with a bottom of india rubber; eggs underneath are pressed up so that the eggs come in contact with the rubber cloth. The water is caused to circulate by means of a stove placed at one side.

MINASI'S INCUBATOR, contemporary with Cantelo, was a more elaborate affair. A boiler heated by a naptha or spirit lamp, all enclosed in an upright box, communicated heat to a reservoir or tank of water. The under side of this tank was corrugated. so as to support by the aid of wires, a series of small narrow sand-bags. against which the eggs were pressed by springs under the drawers.

This device was abandoned and a series of tubes substituted. through which the hot water circulated. The eggs were placed on these tubes, which were close enough together to prevent the eggs from falling through. The machine was too elaborate to become popular.

ADRIEN & TRIOCHE'S INCUBATOR.

Similar to Minasi's first machine, was one constructed by Adrien & Trioche, at Van Girard, France, in 1848. This also had the rubber cloth bottom to the water tank, supported on rods of wood. A sheet iron cylinder, heated by a charcoal fire, supplied the tank with hot water. The top of the tank was covered with sand, so as to retain the heat. This Incubator was ten feet long by three and a half wide. The drawers were in two ranges, placed back to back; the bottoms of the drawers were of perforated tin or fine wire cloth, covered with bran to keep the eggs level. The machine

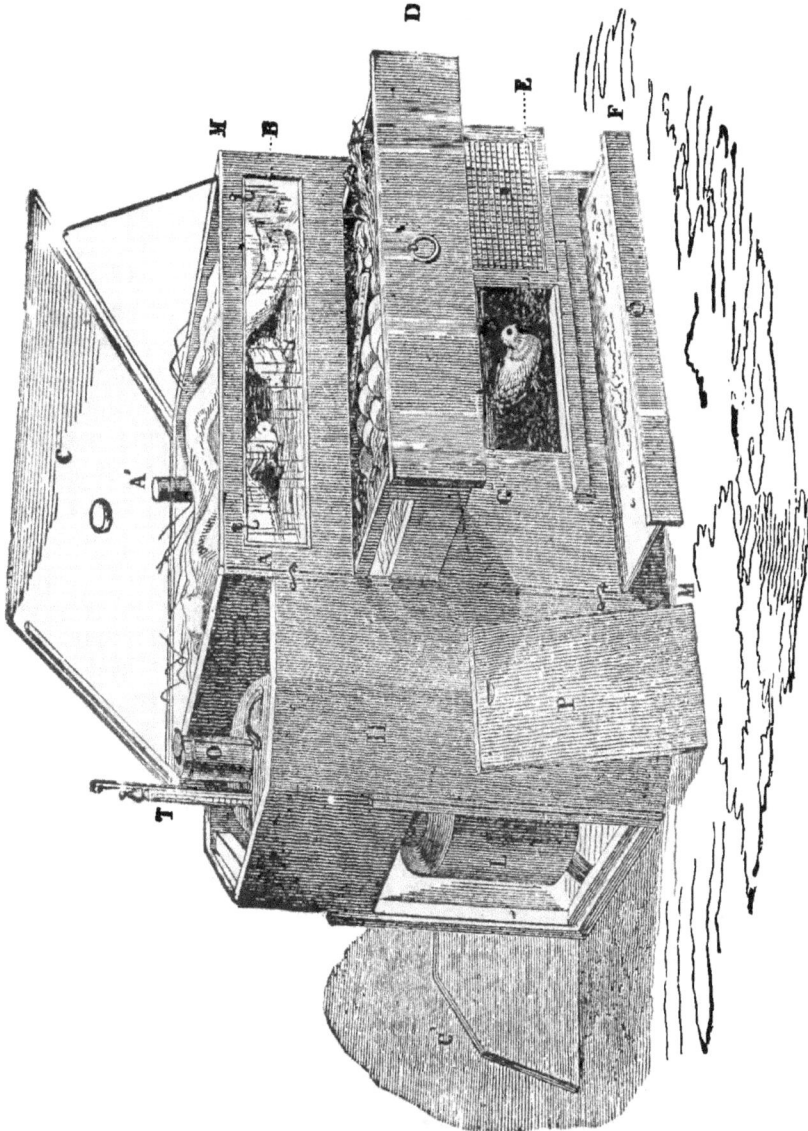

FIG. 52.—VALLEE'S EGG HATCHER.

held fifteen hundred eggs, from which the inventors claimed a result of twelve hundred chicks. It required attention every four hours.

CARBONNIER'S INCUBATOR

which appeared about this time, was a very simple affair. Fig. 50

represents the Incubator, with the drawer containing the eggs, partly drawn out. Fig. 51 shows a section of the same, in which —A, is the zinc case for water—B, Thermometer—C, Non-conducting filling—D, Drawer, with eggs, and E, the Lamp.

The upper part of the box contains a zinc reservoir, with a space left, as shown in the drawing, for the introduction of the lamp, and a small tube passing through the top of the box, which serves for filling it with water, and also for holding a thermometer, which, plunged into the water below, indicates the temperature. Thermometer tubes may be obtained and held in position continually by inserting through a perforated cork of the proper size; the temperature of the water may then be seen at a glance. The drawer for the eggs is immediately beneath the reservoir; it is provided with two small holes for ventilation, and holds about forty eggs. A small thermometer is also kept in the drawer, to indicate the temperature of the air surrounding the egg. A space is left around the reservoir and on

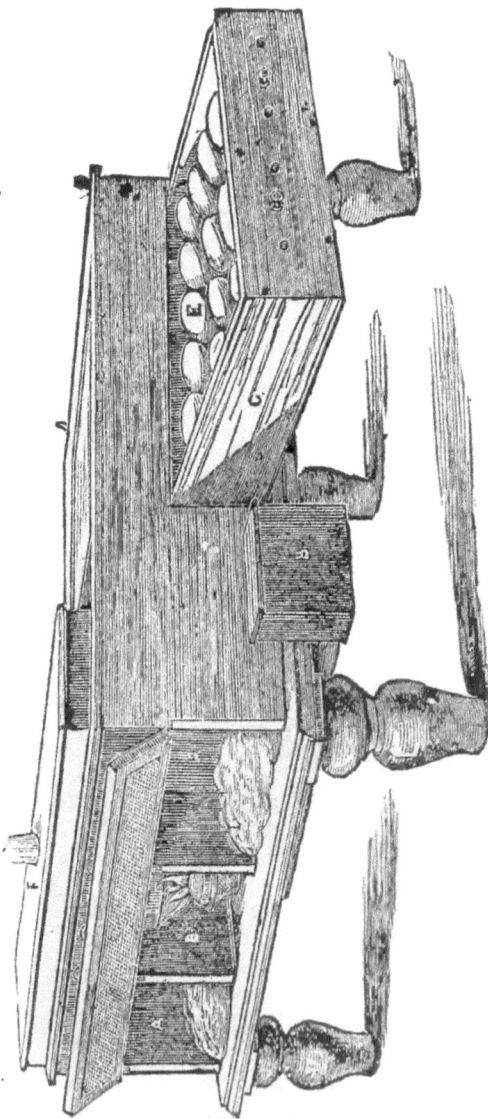

FIG. 53.—BRINDLEY'S INCUBATOR.

three sides of the drawer, for a filling of sawdust or other non-conducting material. A flat tin lamp, with two round wicks, is used by the inventor, but a properly constructed kerosene burner would answer the purpose. A little soft hay is spread in the bottom of the drawer; the eggs are put in; it is then closed and warmed by the water above. The temperature of the water is kept at 122°, or enough higher or lower to keep the eggs at 104° to 106°. Once or twice each day the drawer is opened, and the eggs turned and left for a quarter of an hour in the open air before replacing.

VALLEE'S EGG-HATCHER.

This machine is shown at Fig. 52. It is probably one of the most successful of the many French inventions. One of them was on exhibition for a long time at the Museum of Natural History, in Paris, and gave a very fair percentage in hatching.

In the illustration L is a cylinder, with a flue O, through which passes the flame of the lamp which supplies the heat. This cylinder is, in fact, the *boiler*, connected by a tube to a tank or reservoir *over* the egg-drawer D. From the further end of the case a pipe G, returning under the drawer, carries the water to the boiler again, and thus keeping up a circulation. A is a ventilating tube centre of from the D, and C drawer er of the the covB, where nursery young

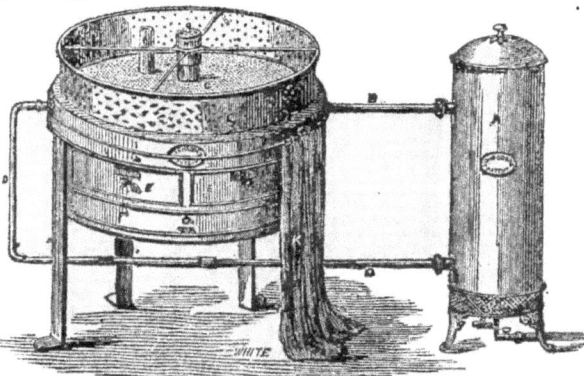

FIG. 54.—SCHRÖDER'S INCUBATOR.

chicks are placed after hatching. E is the sliding door of another apartment, also for the chicks of a few days old, and F a loose bottom to be drawn out and cleaned daily. H is a wooden drum or box enclosing the boiler, and shielding it from sudden drafts of air, which might extinguish the lamp. T is the tube through which the boiler and reservoir are filled with water. The thermometer is laid in the drawer with the eggs.

This machine is still in use in some parts of France. Within a few years a mercurial regulating attachment has been affixed to it, which has much increased its efficiency. It is, however, one of the class of machines which cannot be safely left many hours without danger of overheating.

A number of Vallee's machines have found their way to America. Dr. Preterre, of New York City, had one, with which he was very successful. His apparatus was connected with an electrical bell, which signalled an alarm whenever the heat exceeded the proper limit. It was exhibited by the Doctor a number of times before the American Institute, and also at the poultry shows in N. Y. City, and in 1870 the N. Y. State Poultry Society awarded him a gold medal.

BRINDLEY'S INCUBATOR.

This machine is represented at Fig. 53. *F* is a copper boiler, heated by either gas or lamp *B*, which is furnished with a reservoir to ensure an even height of oil and a steady flame. Connected with the boiler are a number of iron pipes, arranged horizontally between two glass plates, through which the hot water circulates; the space between these glass plates being enclosed on all sides, forms a hot air chamber. Under the lower plate slides a drawer *C*, lined with felting, which contains the eggs. On each side of the lamp, at *A*, are small chambers in which the chicks are temporarily placed after being removed from the egg drawer. The hot air chamber is provided with a safety valve, acted upon by the expansion of mercury in a balanced tube. This valve and regulator was very similar, if not the same, as that used by M. Vallee.

FIG. 55.—THE MOLL PITCHER INCUBATOR.

COL. STUART WORTLEY'S INCUBATOR.

About the same time as the one just described appeared Col. Stuart Wortley's machine. It was described as a long, low, saddle-backed boiler, over which was a steam dome and escape valve. At a short distance, and connected with the boiler by a pipe, was a reservoir, thus ensuring a uniform height of water. A glass gauge was placed alongside the dome to indicate the height of the water. A tank, the same length and height of the boiler, is connected with it by pipes of perhaps two feet long. These pipes pass through padded holes into the upper part of the egg chamber, the boiler being outside. The degree of heat is regulated by sliding the tank and pipes in or out, as may be needed, thus increasing or decreasing the radiating surface. A lamp or gas jet was placed under the saddle of the boiler, and the water kept constantly boiling, by which it was supposed the heat in the drawer would always be uniform.

This machine, while possessing some good points, was too costly, as well as too complicated, to receive much attention.

SCHRÖDER'S INCUBATOR.

Fig. 54 gives a view of this machine, which was the invention of of the manager of the National Poultry Company, Bromley, England.

A separate boiler A, heated by a lamp or gas jet, is connected by the tube B to the hot water tank C, through which the water circulates, and passing out at the opposite side, is conveyed through the pipe D back to the boiler. The tank has an open tube I, in which

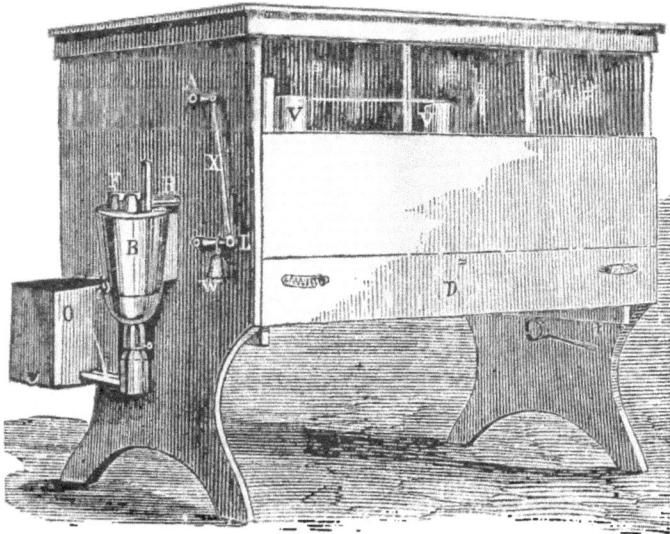

FIG. 56.—HALSTED'S AUTOMATIC INCUBATOR.

a thermometer is suspended. A ventilating flue H is also provided. Under the tank are the egg drawers EE, which are the shape of a quadrant of a circle. The bottom of the drawers are of perforated metal or fine wire cloth, and kept partly filled with chaff. Under the drawers is a tank of cold water. The space G above the hot water tank is surrounded by perforated zinc or wire cloth, and partly filled with sand; this serves the double service of preserving the heat and a warm chamber for the newly hatched chickens. Curtains are arranged to be drawn around the apparatus to shield it from cold drafts.

The mode of testing the heat in the drawer through the temperature of the water was a serious obstacle to the success of this machine, and the lack of a regulating apparatus a still more serious defect; still it had, in careful hands, a fair measure of success. It was found necessary to interchange the egg drawers frequently, as the heat varied in different parts of the egg chamber.

This machine was also too elaborate and costly to meet general favor.

About 1866 the subject of Incubators received a new impetus in America. Among the first which claimed public notice at this period was

THE MOLL PITCHER INCUBATOR.

As first constructed and shown in New York in the fall of 1867, this was made of zinc and some-what similar in construct-ion to Val-lee's ma-chine.

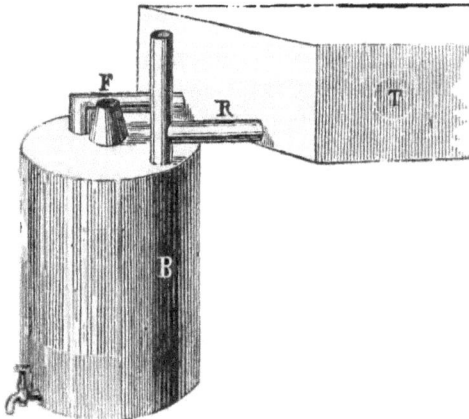

FIG. 57.

It was after-ward recon-structed, and made in form like our illus-tration, Fig. 55. The case is of wood, divided into three com-partments — the lower one for the lamp; the second, a hot air cham-ber in which the eggs are placed in tiers, and the third or up-per one, arranged as an artificial mother for the chicks. A ventilating flue from the egg chamber is seen above. The artificial mother is ventilated by the sliding valves near the top, as shown in the cut. There is no regulating apparatus, the ma-chine re-quiring hourly at-tention and the turning of the lamp up or down, as the heat gets too high or too low.

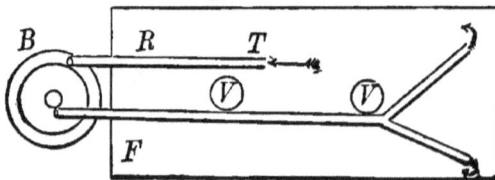

FIG. 58.

THE AMERICAN INCUBATOR

was shown the following year at the exhibition of the Pennsylvania State Poultry Society, in Philadelphia. It was a zinc tank, with a boiler under the centre and opening into the tank, heated by a kerosene lamp. On each side of the boiler was an egg drawer, and also two more over the tank, with an open space of two inches between them. Under the last two, and between them and the tank, were interposed shields of wood placed on an incline, and opening in the middle into the open space between the drawers.

Over these drawers was another shield, opening at each end into the brooding apartment above. The current of hot air generated by the tank passed under the lower shields and upwards between the drawers, over the eggs, and thence into the brooding apartment. The two lower drawers were ventilated by flues or tubes passing through the water tank, and opening into the open space under the lower shields.

This incubator was the invention of the writer. In practice it was found that the upper drawers were useless, as the heat would invariably be from ten to fifteen degrees higher than the lower ones.

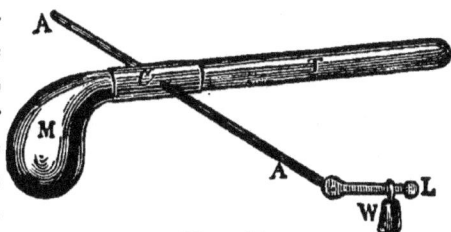

FIG. 59.

HALSTED'S AUTOMATIC INCUBATOR,

illustrated at Fig. 56, appeared the following year.

This shows view of the front and one end of the machine. B is the boiler under which is the lamp which supplies the heat, and which is filled with oil from the reservoir O. D is the egg drawer, above which is the water tank (see T, Fig. 57) which supplies the heat. This tank is elevated two inches above the top of the drawer, so as to give room for the regulator below. It is supported by battens nailed or screwed fast to the ends and sides.

Fig. 57 shows the boiler B and a section of the tank T, with the tubes F, which supplies the tank with hot water, and R, which returns it to the boiler again. Also stop cock through which the water is drawn off. Fig. 58 is a vertical view of the tank. The boiler and tube or flue F taking the water from the

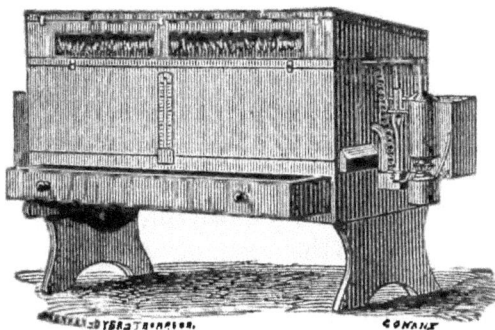

FIG. 60.—GRAVES' INCUBATOR.

boiler and discharging it at the extreme end of the tank, and the return flue R bringing it back again to the boiler to be reheated. V V are two open tubes passing through the tank, acting the double purpose of stays to keep the tank from bulging, and as ventilating flues for the egg drawer. On the upper ends of these flues are fitted valves (like the damper of a stove-pipe), connected and fastened to a rod which passes through the end of the case, and to which is fastened the lever *A* (Fig. 56). The drawer *D* is of wood, inside of which is a zinc or galvanized iron tray just fitting the drawer and tacked to the sides all around; this tray should be about two inches deep. Inside of this is the rack on which to place the eggs; this rack is formed of strips of zinc or iron, each bent like an inverted V, thus ΛΛΛ, and placed one and three-quarter inches apart from centres; these inverted Vs are laid crosswise on three or four supporting tubes or similar strips, to which they are soldered, and the whole rack is then surrounded by a strip of iron one and a half inches wide, to which the ends of these Vs and their supports are also soldered. This rack, being then supplied with handles, can be lifted out of the drawer with the eggs on, whenever

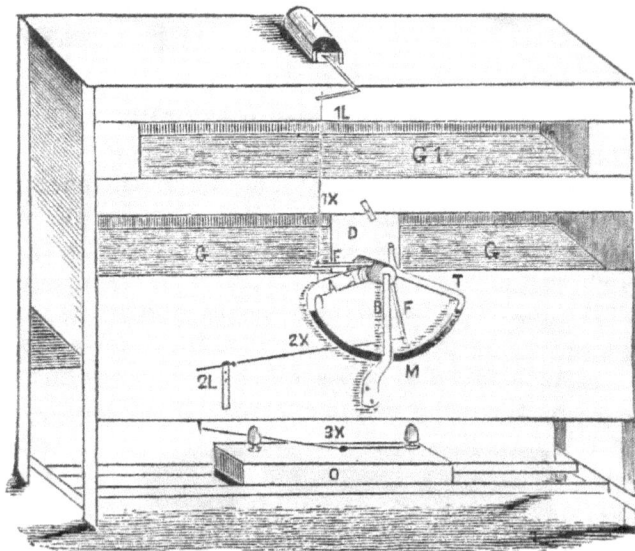

FIG. 61.—HALSTED'S HOT-AIR INCUBATOR.

desired. Over this rack is laid a piece of flannel blanket or other woolen goods, allowing it to dip down into the water, (with which the tray should be filled to the depth of one-quarter or one-half an inch), at every third opening, or oftener if more moisture is desired. The eggs are placed on this blanket. The object of this is to supply a moderate amount of moisture to the bottom of the eggs.

In the front of the drawer, about where is seen the letter *D*, should be a small hole through which to insert the glass tube of a thermometer, the bowl being inside. On the tube should be marks corresponding to 100° and 105°. By this thermometer the regulator is adjusted.

Fig. 59 shows the regulator. This is a glass tube *T*, six or eight inches long, with a bore of five-sixteenths or three-eighths inch diameter, one end of which has a bulb M. This bulb is filled with mercury, sealed hermetically. This had best be done by a manufacturer of thermometers or a glass blower. A brass cylinder C, one and one-half inches long, just fitted over the tube T. On the opposite sides of this are soldered firmly the ends of two pieces of wire AA. On the long end of this wire is fitted a lever L, on which is hung a weight W. This regulator is suspended under the hot water tank and above the eggs by the wires AA, the lever end passing through the case, so that the lever and weight are outside the Incubator, as shown in Fig. 56. Before putting it in place the tube is firmly fastened within the cylinder C, by running in a little plaster.

Returning now to Fig. 56, we notice that the lever A is connected by the rod X with the lever of the regulator L, and the two valves operated by the rod A above mentioned, when open, carry off the hot air from the egg drawer. These tubes may be lengthened and brought through the top of the case if desirable. To regulate the Incubator, when the heat has reached the desired point, balance the regulator with the aid of the weight. As the heat

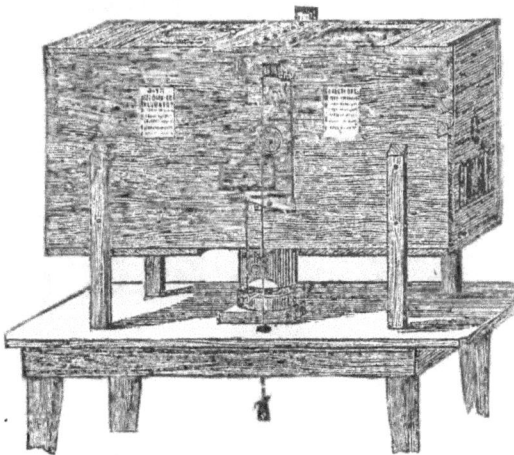

FIG. 62.—DAY'S AUTOMATIC INCUBATOR.

rises above this point, the mercury expands and flows toward the end of the tube T, thus destroying the balance and opening the valves. The fresh air coming in cools off the drawer; the mercury recedes, and the regulator resuming its position, closes the valves. Care must be taken in hanging the regulator to place a stay or catch under the end of the tube so that it shall not quite assume a horizontal position. This will ensure the return of the mercury into the bulb as the heat decreases.

This Incubator met with very fair success, its objectionable points being the liability of the regulator to get broken, owing to the great weight of the mercury in the bulb, and the oxidation of the mercury in the tube. The wetting of the cloth under the eggs by capillary attraction had to be abandoned, as it spoiled the eggs.

GRAVES' INCUBATOR.

This machine was, as first made, " Halsted's Automatic ", recon-structed with a different regulating apparatus. Fig. 60 gives a representation of it. As will be seen, it bears a resemblance to the one just described. The regulating apparatus consists of a horizontal long glass tube, ending in a bent, U shaped extremity, the end of which extends upwards above the horizontal part several inches. The large tube is inside the egg drawer, and the bent extremity outside, at one end of the machine. The horizontal portion is filled with alcohol; the U shaped part with mercury, on which floats a cork connected by a rod to the lever above, which opens and closes the ventilators. The heat causes the alcohol to expand, and that forces the mercury to rise, carrying the float with it.

The hot water tank, cold water pan, etc., are identical with that shown at Fig. 56.

Later, the Graves machine was entirely changed, and the heating arrangement constructed on the plan of the Hoffman Incubator (see Fig. 49), with heat all around; the same regulating apparatus was retained. This last construction proved a virtual failure, and its manufacture was stopped.

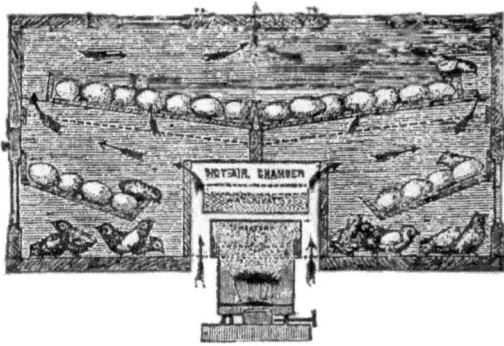

FIG. 63.

HALSTED'S HOT-AIR INCUBATOR.

This machine was the result of an attempt to dispense with the hot water tank and boiler, and carry a current of heated air through the egg chamber from a heated surface of sheet iron or other metal.

Fig. 61 gives an illustration of the outward appearance of the machine. The lamp O was placed directly under a horizontal sheet of iron, which was thinly covered with sand to equalize and retain the heat; above this was a false bottom extending from each end to within half an inch of the middle; on this bottom the egg drawers rested, leaving a space of three inches between them, through which the hot air passed upward. At each end of the machine was a ventilating flue, which extended the whole width and opened into the nursery G1, above which was the ventilator V. Openings on the sides were provided to ensure a current of fresh air passing into the heated chamber.

The regulating apparatus was a long horizontal glass tube, which was filled with alcohol; this was suspended between the drawers, the back end terminating in a pin held by a bracket, and the front in a small neck, which also ended in a pin, which latter was socketed into the bracket B; on one side of this neck was attached the bent glass tube AT, which was filled with mercury as shown by M; attached to this were levers F, F, one of which was connected

FIG. 64.—BAKER'S INCUBATOR.

by the wire $1X$ to the lever of the ventilator $1L$, and the other by the wire $2X$ to the lamp lever $2L$.

The heat in the egg chamber GG caused the alcohol to expand, and that forcing the mercury toward the extremity of the tube, the regulator being hung on pivots, the end T was carried down by the weight of the mercury, and the levers moving with it, opened the ventilators and turned down the lamp flame; as it cooled, the action was reversed.

When first set up this regulating apparatus worked to a charm, but its regularity was of short endurance. The evaporation of the alcohol and the oxidation of the mercury would put it out of order in from five to seven days. The machine itself was not practically worth a rush. It was impossible to keep an even heat in all parts of the drawers, and equally impossible to maintain an even and proper moisture in the air without the closest attention. A hatch of fifty per cent. was a rare occurrence, while the spoiling of the whole batch was the usual result.

DAY'S AUTOMATIC INCUBATOR.

Fig. 62 gives an outside, and Fig. 63 a sectional view of this machine. Like the one just mentioned it is a hot-air machine. The heat is generated by a kerosene stove, over which is a pan of water to keep the air moist, and a hot air chamber, from which the heated air passes over the eggs as shown by the arrows in Fig. 63; the shelves for the eggs are placed on an incline, that the eggs may receive an equal heat; under the lower shelves on each side is an apartment for the young chicks. The regulating apparatus consists of a thermostatic bar, which acts upon a flattened drum; on this is wound a cord which passes over a pulley and is weighted at the other end. The axle of this drum terminates in a crank, to which is attached a wire which operates upon the flame of the lamp, increasing or decreasing it as required. Although seemingly well provided with ventilating holes, the ventilation is bad, and the heat of the drawers varies from ten to twenty degrees. The thermostatic bar is too crudely made to be reliable. The mode of communicating the action of the bar to the lamp though, is admirable from its simplicity and non-liability to derangement.

BAKER'S INCUBATOR,

Illustrated by Fig. 64, was apparently founded on the model of the machine made by Hoffman. It consisted of a hot-air chamber, surrounded on all sides by a tank of water, heated by a boiler placed at one side, the whole enclosed in a wooden case. A battery was attached, and by means of a circuit closer a bell was rung whenever the heat got above or below the proper limit. Later, the electric current was caused to operate upon a magnet, which opened and closed the ventilators. I believe this to have been the first practical application of electricity to the regulation of the Incubator. The ringing of the bell as an alarm, had been used both by Dr. Preterre and myself, some years previous to Baker's adaptation of it.

This machine was most excellent in construction, the tank being of copper, and boiler and flues of same material. But the same difficulty crops out in this as in all other machines, where it is attempted to place several tiers of drawers in one egg chamber—the variation of the heat between the upper and lower drawers must be so great, that either top or bottom must be useless, or else the ventilation must be almost entirely stopped. The machine therefor was only a partial success.

MYERS' INCUBATOR.

This was an adaptation of the principle used by Cantelo, and by Adrien & Trioche. A large box eight feet long by three or four wide, was fitted with a tank made of india rubber sheeting, supported on wires placed two inches apart. Under this were the

FIG. 65.—THE NATIONAL INCUBATOR.

drawers, also fitted with rods placed two inches apart, and so arranged that the eggs came between the supporting wires of the tank, when the drawer was fitted into its place. The drawers were in two ranges placed back to back, as in A. & T.'s machine. The hot water was supplied from a stove and boiler standing a few feet away, heated by a coal fire. A regulating attachment was provided, which was said to act upon the damper of the stove, opening or closing the draft and thus increasing or decreasing the heat.

Over the tank was placed a zinc or sheet iron covering on which sand was strewn. This was surrounded by a wire-gauze railing a few inches high, and was used as a nursery for the newly hatched chickens.

The inventor had very good success with this machine; taking eggs from the farmers to hatch as well as hatching for himself. It required though, very close attention. I do not·know that any were ever made for sale.

THE NATIONAL INCUBATOR

is illustrated by Fig. 65. It is a hot-air machine, and as will be seen, is formed of many parts. First. is the stove on the right, which is heated by kerosene, supplied from an inverted can (seen under the machine proper) which discharges the oil into a small cup, whence it is conveyed to the lamp through a small tin tube. As the cup is open, there must be a great deal of evaporation as well as odor from the oil. The stove is simply two drums, one inside the other, the space between them packed with non-conducting material. The Incubator proper is a circular box, (like a large cheese box) made of wood, covered in some cases with tarred paper, and supported on a frame as shown in the cut.

The drawers for the eggs are semi-quadrant in shape, inclining from the outer edge towards the centre. The bottoms of these are covered with sand, on which the eggs rest. These drawers all rest on a revolving bottom, which in turning brings each drawer in turn before the door. The hot air generated by the stove, is carried into the machine through the connecting pipe, and by extending the pipe inside, is discharged into the centre of the egg chamber. Directly under this point is a vial of mercury, into which is placed one end of a platinum wire, this is fastened to a connecting copper wire, which leads to the battery; the return wire is also tipped with platinum, and is so set that when the heat causes the mercury to rise to a certain degree, it touches the platinum wire and the electric circuit is complete. When this is done, an electric bell commences ringing and ceases not until the heat recedes again. We do not know of any other regulating (?) apparatus about it; (although it is claimed that the heat is shut off from the egg chamber when the temperature gets too high:) the maintaining of the proper heat depending on the care and watchfulness of the attendant.

Living in the same room with the machine, and sleeping beside it, the inventors claim to have hatched some wonderful percentages. They certainly have made some most excellent exhibitions at the different shows throughout the country; and if they could succeed in devising some less odor absorbing material for the bottom of their egg drawers, or change the present material oftener, hundreds of visitors would appreciate the innovation.

HYDE'S INCUBATOR.

This was designed and constructed for the pleasure and use of the inventor, but as several of them have been made and found their way into other hands, it is well to give a short description of it.

The heating arrangement is similar to Vallee's; the machine stands on four legs like a table; on the opposite end from the lamp is placed the regulating apparatus. This consists of a short and long arm revolving on the axis of a clock movement, and catching alternately on the notched arc of one end of a lever, the other end of which is moved by a thermostatic bar in the egg chamber of the machine. Attached to this axis is also a crank, from which a very

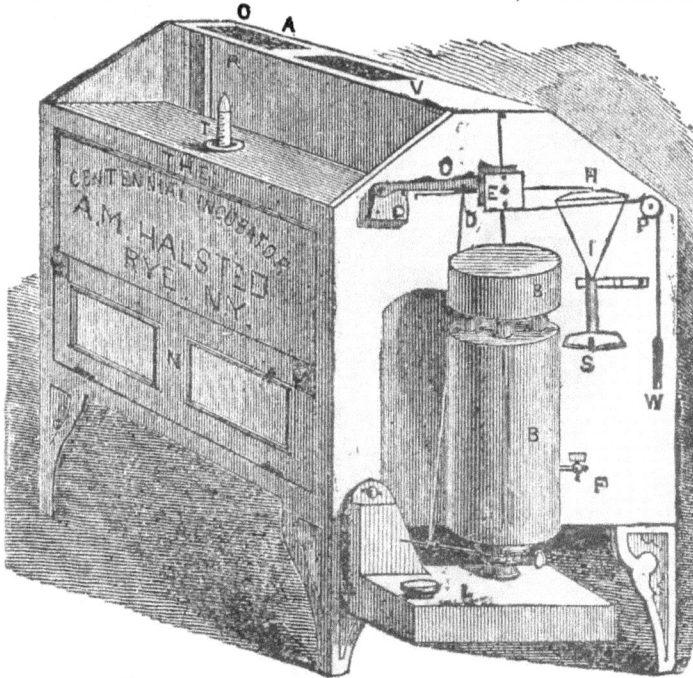

FIG. 66.—THE CENTENNIAL INCUBATOR.

light wooden rod connects with the lamp, raising or lowering the flame as needed. Under the egg drawer are placed sponges saturated with water, to give the necessary moisture to the air.

THE CENTENNIAL INCUBATOR.

This machine, illustrated at Fig. 66, is the outgrowth of a great many experimental machines, two of which are illustrated at figures 56 and 61.

The principles embodied in the present Incubator were first combined and shown at the International Exhibition at Philadelphia in 1876, from which it received its name. At that time the case of the Incubator was of wood, and the boiler, tank and lamp only of metal. The following year the machine was remodelled, and constructed entirely of galvanized iron, save the heating arrangements, which are of copper.

The growth of this machine may really said to have begun in 1865, at which time the writer began his first experiments in Artificial Incubation, but it was not until 1878 that The Centennial ap-

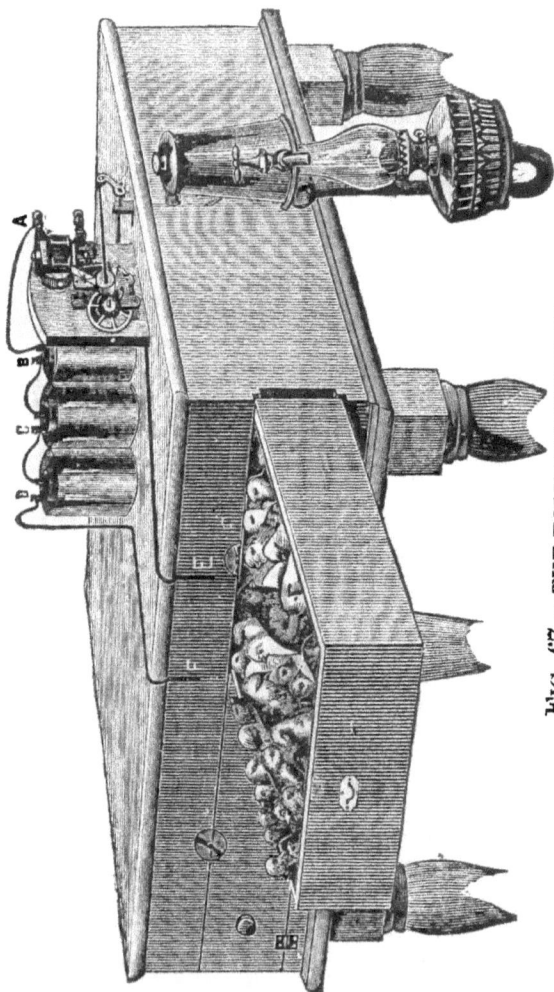

FIG. 67.—THE ECLIPSE INCUBATOR.

peared in its present form and took its hard-earned position as the Standard Incubator of America, if not of the world.

As nearly one hundred of them are in use on the ostrich farms of South Africa, and fully fifty more distributed in England, Germany, Australia, Cuba, Mexico and Brazil, the title is not inappropriate.

The Incubator consists of a copper boiler and tank, the boiler encased with an outer jacket of galvanized iron, and the space between packed with mineral wool. The tank is enclosed by a case of galvanized iron, outside of which is still another case of same metal, with intervening space also packed with non-conducting material; this space is from two to three inches thick. Inside the tank is a system of tubes which bring the water from the boiler and so distribute it that the outer edges of the tank are the hottest; the water, as it cools, is taken up by the return flues and conveyed back to the boiler, to be again heated. The heat is generated by a lamp with a kerosene burner, such as is used on a large-sized table lamp. The chimney is of metal, and passes through the water, thereby preventing any waste of heat by radiation.

The egg chamber is under the tank, the drawers so constructed that the eggs are held by parallel rods in position where placed; under the egg drawers is a cold water pan, to keep the bottom of the eggs cool and moist; the bottom of this water pan comes in direct contact with the outer air, thus ensuring the water from getting heated. In the egg chamber are also water vessels to provide moisture for the eggs. Over the eggs, and directly under the tank, is the thermostatic or regulating-bar. This bar is of composite character, and is the result of many years experimenting. Many of them have stood the test of three consecutive seasons, which is a very thorough test of their reliability and lasting qualities. This bar is so attached to the machine that it can be regulated from the outside, and the free end (the point of which is shown at S, Fig. 66) adjusted to the one-hundredth part of an inch if desired.

As we shall need frequent reference to the cut in describing The Centennial, we give the parts here:

A—Tube through which the tank and boiler are filled.
B—Boiler in two parts: B—the boiler proper; and B'—the boiler head.
C—Lamp-lever.
D—Connecting rod between lamp-lever and lamp-trip.
E—Reel or drum.
F—Faucet to draw off water.
H—Escape.
I—Escape-lever.
L—Lamp.
N—Door of egg chamber.
O—Vent tube.
P—Pulley.
R—Reservoir.
S—Wire end of the regulator-bar projecting through slotted hole.
T—Thermometer.
V—Ventilator box or flue.
W—Weight.

A ventilating flue extends from the egg chamber through the tank opening at V; another circular hole through the tank allows the thermometer to extend down into the egg chamber, but leaving the scale above, so the temperature of the eggs may be seen at a glance.

The regulating apparatus consists of a reel furnished with four arms, whose bent ends are caught and released alternately by the curved end of the escape-lever, the lower end of which is operated

FIG. 68.—CORBETT'S APPARATUS.

by the wire point of the regulator-bar; a cord or thread wound on the reel passes over the pulley, and has a weight attached which gives the power to the reel.

Another lever C is operated by two pins on the back of the reel; this lever is connected by the rod D with the lamp trip. The shaft of the reel passes through the side of the case into the ventilator flue, and is there attached to a revolving ventilator.

As the thermostatic bar moves forward or backward with the heat, the opening in the escape-lever comes over one of the bent ends of the arms of the reel, allowing it to pass through, which it does by the force of the weight, the next arm catching on the opposite side of the opening; this opens the ventilator, and at the same time turns down the flame of the lamp. As the drawer cools, the escape-lever is moved by the thermostatic bar in the reverse direction, the next arm is released, and the reel makes another quarter revolution, closing the ventilator and turning up the flame of the lamp. The action is repeated until the weight runs down. The lamp is hinged to the side of the Incubator so that it is impossible to upset it, unless first removing it from the machine.

As will be seen, this Incubator is very simple, very complete, easily regulated, not liable to get out of order, and, being all of iron and copper, will last for many years. It is also handsomely grained in oak, japanned and kiln-dried; added to this is its automatic action, being so positively self-regulating that it may be (and has been) left for sixty hours without attention, and the temperature in the egg chamber not varied over three degrees during that time.

Another feature worthy of consideration is the fact that purchasers of the machine have better success in hatching than the inventor himself, their averages being larger, and in five instances every fertile egg was hatched, or one hundred per cent.

Letters patent were granted for the Centennial Incubator in July, 1880. For further particulars send for illustrated circular to A. M. Halsted, Rye, N. Y.

THE ECLIPSE SELF-REGULATING INCUBATOR—FIG. 67.

In this Incubator the heat is radiated from a galvanized iron tank placed above the eggs, in which is a constant and regular circulation of water. The heating is done by a kerosene oil lamp, which consumes less than a quart of oil in twenty-four hours. By no possibility can the fumes of the lamp enter the Incubator, or reach the eggs. The Incubator is thoroughly ventilated by six pipes beneath the egg drawers, so the bottoms of the eggs have a constant supply of cool, fresh air.

The entire bottom of the Incubator is covered with water, to furnish moisture to the eggs by evaporation. In the Incubators we now build this water can be changed in a moment, without

disturbing the eggs or even opening the egg chamber, and either
hot or cool water can be supplied, according to the fancy of the
operator.

The electric battery of the Eclipse is a very simple one; each

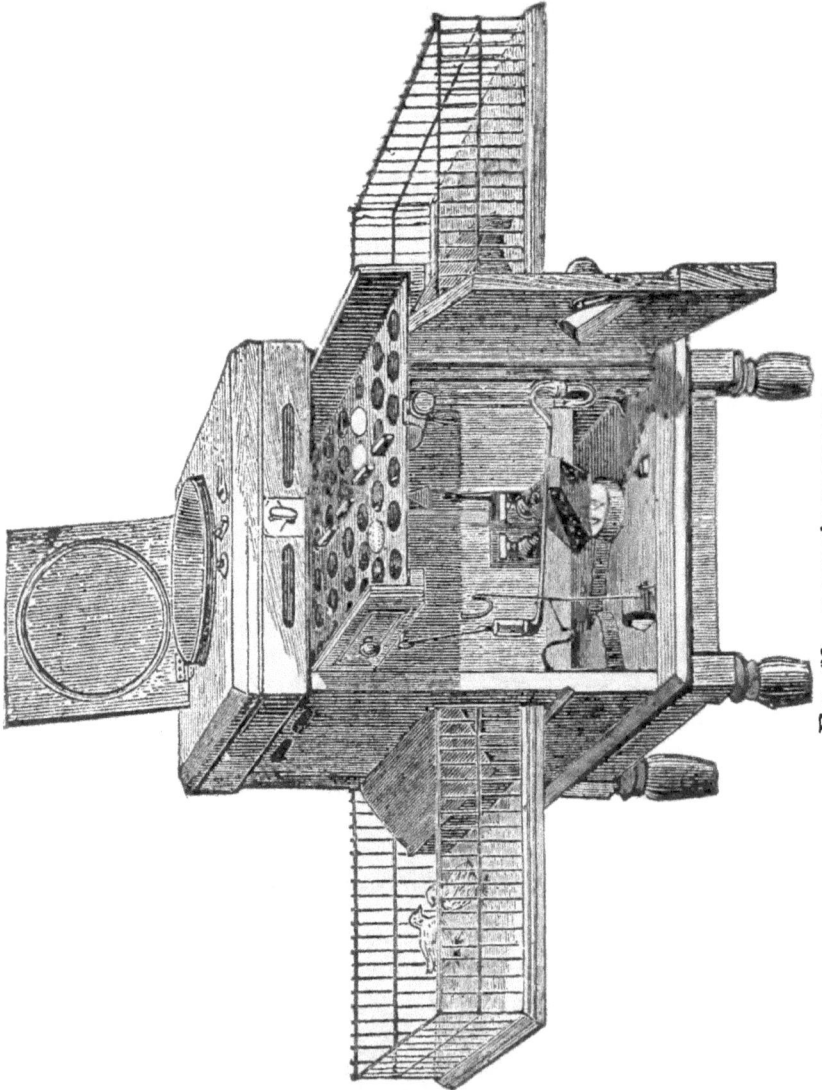

FIG. 69.—BOYLE'S INCUBATOR.

purchaser puts it together himself, so he cannot fail to fully understand it; nothing about it is covered—all is in plain sight—and
if it should become exhausted it need not be returned to the manufacturers for a new one, as the purchaser can set it right again
himself in about one minute.

The Eclipse is put together by dovetailing every joint and corner, so it cannot warp, and makes a perfectly air-tight joint.   As

the machine for making this joint costs three hundred dollars, it is not at all likely that any other manufacturer will ever use it. We furnish the Incubators very handsomely, so they can be kept in the dwelling house; it is much more convenient there, especially if it is to be managed at all by women or children.

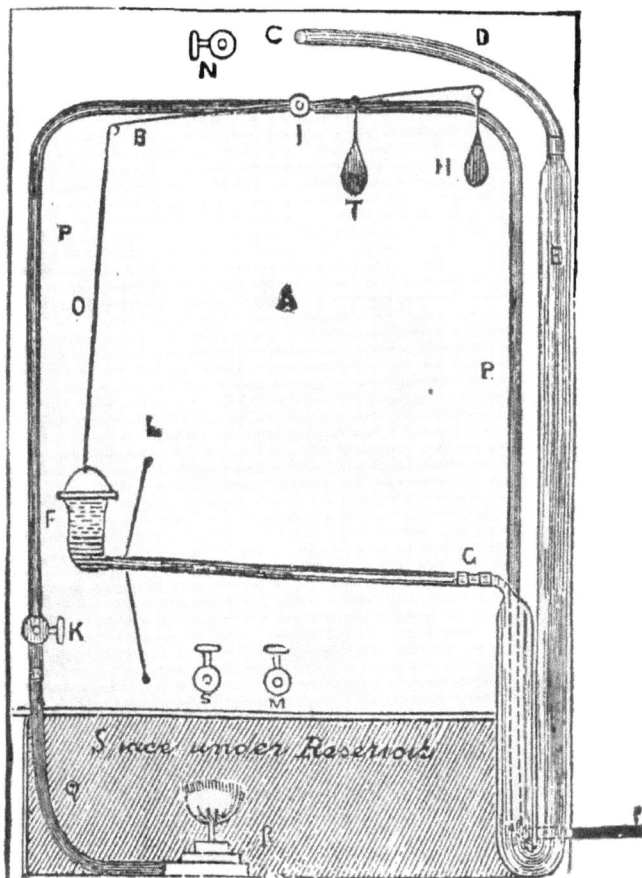

FIG. 70.

The Eclipse is manufactured in Waltham, Mass., and circulars and prices will be sent to any one upon application.

WESTON'S INCUBATOR.

This Incubator had the appearance, outwardly, of an old-fashioned kitchen safe. It stood upright, elevated about two feet from the floor on legs like a table. A tank in the top was supplied from a boiler suspended underneath the case, by two connecting pipes, in each of which were stop-cocks. The method of regulating the heat was turning these stop-cocks, giving a faster or slower circulation to the water, as required. This, of course, had to be done by hand.

The inventor says of it: "The heat is generated by use of coal oil lamps; the heat cannot run too high, and is so arranged that the condensation of steam regulates it, while at the same time it furnishes the requisite moisture to the eggs. It requires no care, except to occasionally fill the lamp with oil". He further says: "After several years of patient study, I have so far perfected the apparatus as to feel it cannot be improved".

As the water in the boiler is never supposed to come to the boiling point, we are at a loss to know from where he gets his steam to condense. The machine, however, like dozens of others, has dropped out of notice.

CORBETT'S APPARATUS appeared about this time. It is simply a modification of Reaumur's method of hatching by means of the heat generated by horse manure. Fig. 68 gives a front view of the apparatus, with the door let down. It is, as appears, a barrel-shaped box with a circular tray or trays to hold the eggs; fitted with a slide, over this is a flue, to give the neces-

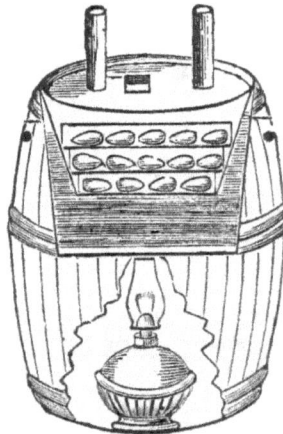

FIG. 71.—JAQUES' INCUBATOR.

sary ventilation. The box, when closed, is packed in horse manure, leaving only the ventilating flue uncovered. The heat of the apparatus is tested by a thermometer, through the ventilator, and kept as near right as possible by opening or closing the slide. The same apparatus is used as an Artificial Mother, by putting in a circular board lined with felting or sheepskin, and elevating it a few inches above the bottom of the box. The front is left open, as shown in the cut.

Mr. Corbett, we believe, owes his first ideas of this system of incubation to Dr. Preterre, of New York City, by whom he was employed, and under whose direction he tried artificial hatching by Reaumur's method on the Doctor's farm in New Jersey. The Doctor stated to the writer that he found the process uncertain and unsatisfactory, and gave it up in favor of the "Vallee" machine, which he had previously used.

Mr. C. carried on hatching operations for a time at Hicksville, L. I., but abandoned them after one or two years experience.

BOYLE'S INCUBATOR.

This is an English invention, and one which bid fair to make a success, but for some reason the results with it, so far as we know, have not been as satisfactory as at first promised.

Fig. 69 gives a front view of the machine, and Fig. 70 a plan of the regulating apparatus.

The machine is about two feet square and three high. Through the open door, which covers about half the entire front when closed, is seen the boiler, with stop-cocks for drawing off the water; the lamp, with its attachment, and the regulating apparatus. Above this are two half doors, which open at right angles to the face of the case, on the under end of which are two projecting battens, upon which the egg drawer slides out.

This egg drawer is a tray, with holes cut through large enough to allow the egg to drop about one-third through. This arrangement is designed to cool the bottom of the egg, in imitation of the natural process when the hen makes her nest on the ground.

Upon the tray are small cups, which are kept filled with water to keep the air moist.

Attached to each side of the machine are two projecting shelves, edged with a low railing, which serve as a run for the chicks for a few days after hatching. Next the boiler, and opening into the runs, are two inclined boards, lined with fleece or some soft material, under which the chicks nestle, and which are, in fact, small-sized Artificial Mothers.

On the top of the case is a circular opening covered with glass, through which the eggs may be seen and the temperature noted. This is further protected by a hinged cover.

Referring to Fig. 70, the parts are designated thus:

A—Front of case.	L—Compensating spring.
B—Balance of lever.	M—Supply-cock of boiler.
C—Connecting pipe.	N—Air-vent of boiler.
D—India-rubber tube.	O—Connecting wire.
E—Glass U tube.	P—Connection with gas supply.
F—Bowl for mercury.	Q—India-rubber connection with flame.
G—Elastic joint.	
H—Balance joint.	R—Gas jet.
I —Pivot gas-cock.	S—Exhaust-cock of boiler.
K—Stop-cock (gas).	T—Movable weight.

The regulator works by the expansion of water and the weight of mercury. When the boiler is filled with water and heated to the proper temperature, mercury is poured into the bowl F until it rises sufciently high in the tube E. This tube being connected

by the India-rubber tube D with the boiler, and the water in the
boiler having no other outlet, when the heat rises, and the water
consequently expands, the mercury is forced into the bowl, and
relieves the balance weight$H$, which partly closes the pivot gas-
cock I, and reduces the supply of gas.   As the heat falls, the water
in the boiler contracts, and the reverse action of the mercury increases the flow of gas, and thus creates a larger flame.   The
compensating springs L prevent the bowl rising or falling too
suddenly.

If a lamp is used, instead of gas, the connecting wire is attached
to the mechanism for cutting off the flame.

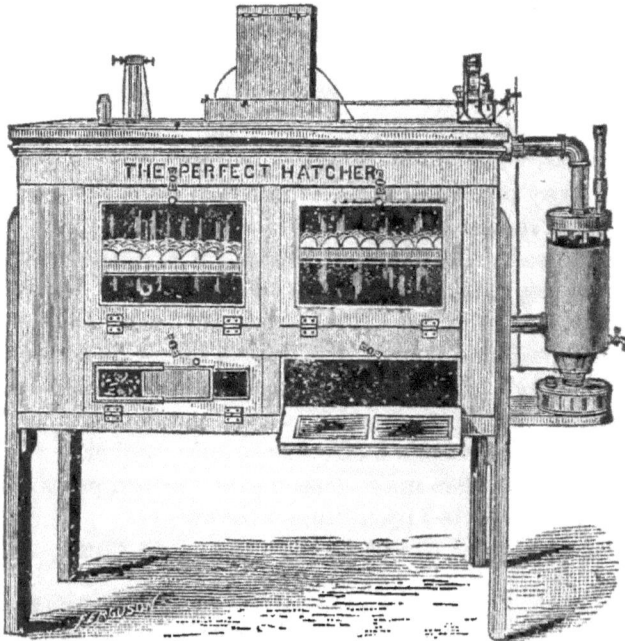

FIG. 72.—THE PERFECT HATCHER.

There is no ventilator connected with the regulator, the regulation of temperature being entirely dependent upon the cutting off
or turning on the source of heat.

Its objections are its complicated arrangements and its great
cost in proportion to its capacity, the Incubator from which our
illustration was taken holding only forty-two eggs, and costing in
England £15—about $75 currency.

JAQUES' INCUBATOR.

This is an American invention, and probably the cheapest apparatus in the Incubator line.   It is shown in Fig. 71.   The inventor says of it:

" The above cut represents an Incubator invented by me for my own use. I have succeeded in securing a chamber in which I can regulate the temperature and hold it at any degree desired. My claim for it is, that it is.the easiest to construct and cheapest to make of any Incubator offered to the public, I do not claim that it will hatch as large a percentage of eggs as those constructed on more scientific principles, yet it will hatch a large percentage if carefully managed by any person of ordinary intelligence. As there is no self-regulating attachment, it will require more care than those that have it. The capacity of one made with a sugar barrel is about one hundred eggs; the cost (excepting the thermometer) about five dollars. I do not manufacture them for sale, but, on receipt of fifty cents, will send my pamphlet, containing full instructions for making and using one ".

CHARLES B. JAQUES, Metuchen, Middlesex Co., New Jersey.

THE PERFECT HATCHER.

Fig. 72 represents a front view of this machine, for which the inventor makes the modest claim that ".it is the only perfectly reliable Incubator in the world". Singular to relate, the inventor seems to be the only one who has made that wonderful discovery.

The Incubator from which our cut was taken is made of wood, and is about five feet long, three wide, four and a half high, and is said to hold three hundred eggs. A tank of water fills the entire upper portion of the case, over the eggs; at the opposite end from the lamp, a return flue takes the water from the tank into the chamber under the eggs, (one door of which is seen open), whence it is carried by a number of smaller pipes, back to the return flue of the boiler. The lamp and boiler are identical with that of the Centennial, (except that the latter has no outer jacket), and were copied from that model. The bottoms of the egg drawers are of fine wire cloth, on which the eggs lie without any intervening substance. These egg-trays are on movable slides, so that they can be raised nearer the tank to compensate for the unevenness of the heat. The lower chamber above spoken of, constitutes a brooder or artificial mother for the chicks when first hatched.

Over the egg drawers, is suspended a rubber rod three or more feet long, fastened at one end to a stationary bracket, and the other end held by a spring acting lengthwise with the rod; this end is attached to the lower end of a lever, which, extending upwards through an opening in the tank, connects with the wires of an electric battery. This battery is used in connection with a clock-work, similar in principle to that of the Eclipse. The expansion of the rod as the heat rises, closes the electric circuit,

which, acting upon the magnet, sets the clock in motion, opening
the ventilator and turning down the light.   As the rod contracts
under the lowering of the heat, the reverse action follows.

Although it is such a "perfect" machine, but very few of them
are in use; the complication of the electric attachment proving a
bar against its popularity; and the unevenness of the heat in the
unnecessarily large egg chamber, preventing as good results in
hatching, as has been attained with a number of other styles of
Incubators.

WATERHOUSE'S INCUBATOR.

This is a Connecticut invention.   In general appearance it is **a**

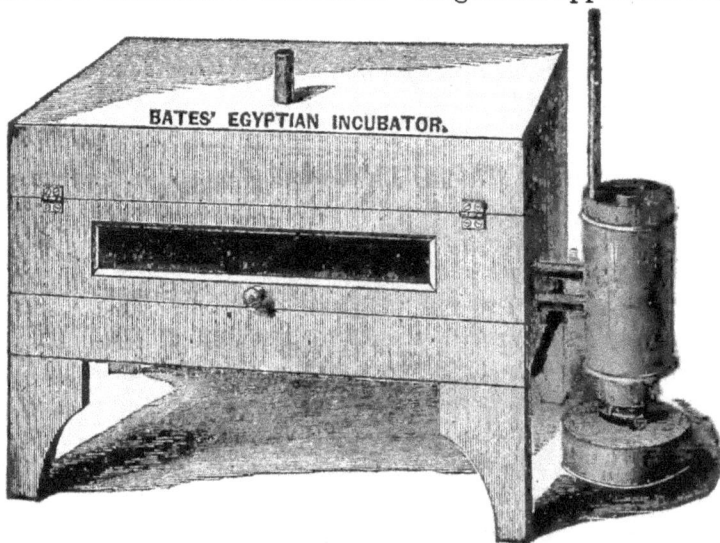

BATES' EGYPTIAN INCUBATOR.

FIG. 73.

little similar to Jaques' Incubator; the sides of the case however,
are built up of either paper or wood.   The eggs rest in a basin-
shaped pan in the centre of the cylindrical case, around the outer
edge of which is a space for the entrance of the hot air from the
lamp chamber.   In the latter is a reservoir of water, to provide
moisture for the air.   Over the egg-basin is a circular glass plate,
through which can be seen the eggs and the thermometers which
lie on top of the eggs.   Above and to one side is a flue, opening
into the outer air; in this flue is a pivoted damper or valve.   The
regulator is a copper rod, firmly fastened at one end and free at
the other, which is connected with the short arm of a nicely adjusted lever, the other arm of which is connected with the damper above mentioned; when the chamber gets too hot, the expansion of the copper rod, is supposed to act on the lever with sufficient force to cause the damper to open, thus drawing off the heat

from the egg chamber. As it cools, the rod contracts and closes the damper.

The strongest objection we see against this Incubator, is that the fumes of the kerosene oil lamp pass directly into the egg chamber, which we hardly think compatible with the health of either eggs or chickens.

BATES' EGYPTIAN INCUBATOR.

If simplicity be of any value in an Incubator, that illustrated by Fig. 73 must be a very valuable apparatus.

A wooden case inside of which is a tank of water; a boiler outside connected by supply and return tubes with the tank; a small size kerosene stove to generate heat; an open space under the tank in which to place the eggs; and a door through which to pass them in or out, comprises all the essential features. There is no regulating apparatus of any kind. Only one thing is necessary to perfect success with it, and that is a watchman to sit alongside of it for three weeks to turn the

FIG. 74.—THE RELIANCE INCUBATOR.

lamp up or down as needed, and shift the eggs around to get uniformity of heat.

MASTERTON'S INCUBATOR,

Comes from the Pacific Coast. Its principal features are the use of a coil of pipe inside the tank, through which the water from the boiler passes before it is discharged into the top portion of the tank; and a revolving egg-tray, by which the eggs are turned without opening the egg chamber. This latter is a drawer or tray, suspended on journals. The eggs are laid on a perforated bottom, through which the heat has free passage; they are covered with a similar material which is securely fastened to the sides of the tray. In turning, all that is necessary is to revolve the drawer half way and the eggs are reversed. The device is patented.

DAVIS' INCUBATOR,

Finds life and recognition at the Patent Office, from Pennsylvania. It has several peculiarities worthy of mention, from the fact that they differ from the stereotyped plans usually seen.

The tank placed in the usual position, is virtually a square basin with a glass bottom and a hinged cover. The hot water passes into one end from the boiler, and is distributed by three pipes in different parts of the tank. Passing out at the other end, it is carried down and under the egg chamber back to the boiler. The drawer is fitted with an inner bottom of pliable material, held up by coiled springs. The eggs are placed on this and the drawer pushed up until the eggs touch the glass bottom of the tank; the coiled wire springs ensure every egg resting against the glass.

In the chamber under the eggs the return pipe is caused to pass to and fro a half dozen or more times, in order to form a brooding apartment for the young chicks.

The apparatus for elevating the drawer consists of four levers, the inner ends of which are connected by cords to a small windlass, which, being turned, depresses the centres and elevates the four corners of the drawer simultaneously. There is no regulating apparatus on the machine.

THE RELIANCE INCUBATOR.

The "Reliance" Incubator was invented by James Dennis, Jr., of Providence, R. I., and patented July 20th, 1880; it is now manufactured by him in that city. It is a first-class machine, and is made in a thorough and workmanlike manner.

This machine consists of two cases, made of the best quality of galvanized iron, the parts of which are both riveted and soldered together. There is one inch of space between the cases, which is filled with the best non-conductor of heat yet discovered. In the upper part of the incubator is a chamber in which is placed a soapstone radiator, that is heated by means of a continuous coil of hot water pipes. Underneath this radiator are the egg drawers, and beneath the egg drawers the evaporating pans are placed. The air for ventilation is taken in through holes punched in the bottom of the machine in such a manner as to materially assist in regulating the heat. The ventilators through the top are so contrived as to thoroughly ventilate every part of the egg drawer without exposing any of the eggs to the direct influence of the outside temperature. The machine is placed on a solid table, made of the best quality of western ash, and, for convenience in manipulating the eggs, it is furnished with a slide shelf on which to rest the drawer. The heat used is obtained from a "Florence" oil stove of a pattern made expressly for this machine. In using a soapstone radiator in preference to a tank of water, the inventor considers he has gained in several particulars, viz.: It is very much less sensitive to changes of temperature; again, he gets a mild, soft heat, nearest like that developed from an animal body, and consequently is best adapted for developing animal life. The ventilation is perfect, as the ventilators are never closed. This is one of the simplest, if not the simplest, machines now in the market, and the inventor claims it will satisfactorily hatch all the eggs that would hatch, under the most favorable circumstances, in the natural way.

It requires very little care, as ten minutes twice each day is amply sufficient to attend to a three hundred egg machine. There is no electricity used in connection with it, and no machinery for the opening and closing of valves, so that there is nothing liable to get out of order. Further information cheerfully furnished upon application to the inventor: address, No. 86 Meeting street, Providence, R. I.

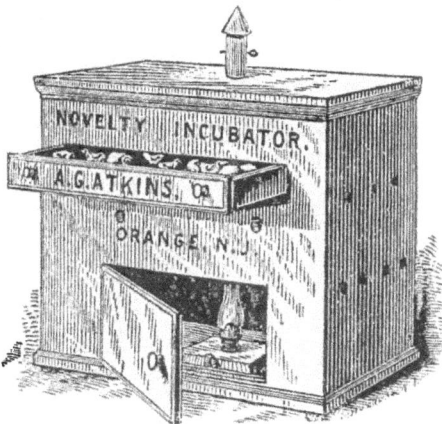

Fig. 75.—THE NOVELTY INCUBATOR.

MEAD'S INCUBATOR consists of the usual tank and boiler, heated by a lamp or gas jet. In general appearance it is somewhat like Carbonnier's, with the lamp and boiler placed outside the case. Over the top is a glass-enclosed space for the newly hatched chicks. It has no specially interesting features that we know of.

THE ONEIDA COMMUNITY HATCHER

was exhibited at the Fair of the New York State Agricultural Society, a few years since, and there hatched out a goodly number of chickens. It was modelled somewhat after the Eclipse, outwardly, and regulated by an electrical battery. It was, however, of very rough and primitive construction, and needed frequent looking after, although a self-regulator.

THE NOVELTY INCUBATOR.

Fig. 75 shows this really novel Incubator. Its heating apparatus consists of two tanks, one above the other, with a space of ten or twelve inches between them; from the top of the lower tank, at the corners, four tubes led into the upper tank, extending nearly to the upper side; from the bottom of this upper tank four more tubes led downwards, discharging into the bottom portion of the lower one.

The lamp was placed under the centre of the lower tank, and from the construction above described a very complete circulation of water was obtained.

The whole was enclosed in a wooden case, with a door through which the egg tray was put in or out, the egg chamber being between the two tanks. Ventilation is provided by holes through the sides of the case and a small flue with a check or stop valve on top. On top of the lower tank, and under the egg drawer, are

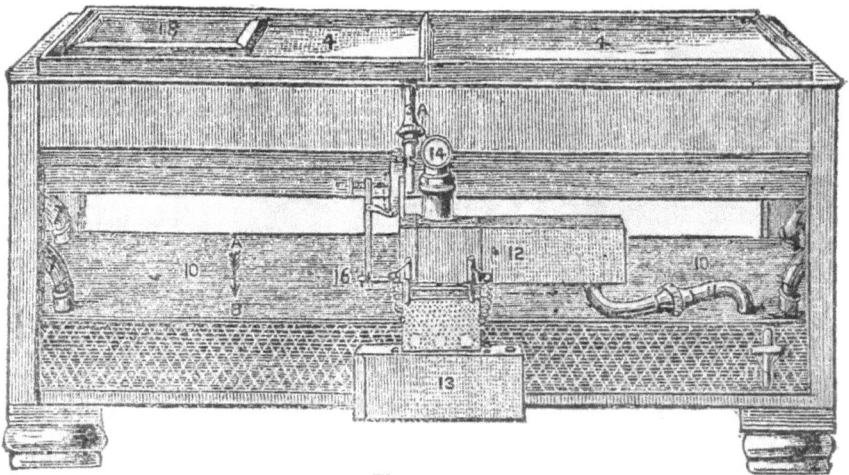

FIG. 76.

placed pans of water, which evaporating, keep the air charged with moisture. No regulator is used, the supposition being that a free draft of air would prevent any overheating of the eggs during the time the machine was left alone, which was placed at four to six hours. In theory this seemed very plausible, but in practice it was found that in warm weather, or in a room of variable temperature, it required almost hourly attention.

CAMPBELL'S EUREKA INCUBATOR

has one feature about it that deserves mention. In general appearance and construction it is a little like "Christy's Hydro" (illustrated further on), but the egg tray or slide has on it a rack, between the partitions of which the eggs are laid. This rack is connected with a clock movement, which is in turn connected with a clock similarly to an alarm attachment; the pointer is set at any specified hour, and when that time arrives the rack is pushed slowly along until the eggs have rolled half over. At the next recurrence of that hour the rack is drawn back, and the eggs turned back to their first position; the eggs thus are turned twice daily, whether the attendant is present or not.

The absence of any regulating apparatus is, however, a serious drawback, and the temptation to neglect the airing of the eggs almost counterbalances any benefit derived from automatically turning them

PENMAN'S INCUBATOR.

This is another English invention, and is said to be a very superior machine. Figs. 76, 77 and 78 show different views, and, by a reference to following description, will give a very excellent idea of the machine.

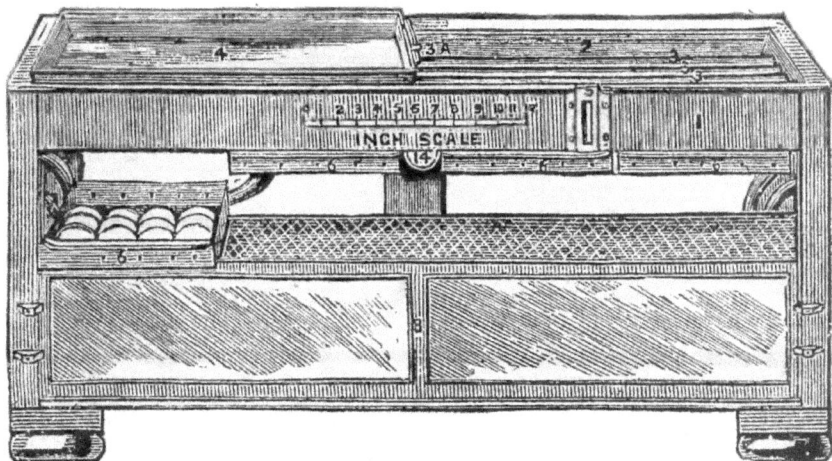

FIG. 77.

1. A wood frame, similar to a table frame, supported by four legs, by a screw and nut arrangement, like a piano.

2. An India-rubber cloth waterproof tray, cistern or tank, made fast to the inside face of the frame (No. 1), and resting on a ledge attached to the same, and on two cross-bars. At each end of the cistern are two outlets (7), communicating with the Artificial Mother underneath, and through which the water flows after having traversed the upper surface of the eggs.

3. Three (less or more) brass or other metal tubes, to contain air, gases, spirits, oils, or any other fluid that can be expanded by heat, extending the whole length of the cistern, and in communication with each other and with the regulator (15, Fig. 76) by means of pipe 3A.

4. Two iron covers for cistern (one out of its position to show interior of cistern) to retain the heat and steam arising therefrom.

These covers can be utilized for the purpose of raising seed, striking cuttings of plants, or growing ferns or other plants. Glass shades or covers of any pattern can be made for them when required.

5. Thermometer (in shape like the letter ⌐) to indicate the correct heat in the egg drawers or trays, which it does by being in contact with the source of heat, and not obtaining it by radiation.

6. Four drawers or trays (one at the left hand out of position) eighteen inches by ten inches, each perforated with a number of small holes half an inch from the bottom for the exit of the foul air engendered in the tray during incubation, and another row of holes near the top for the inlet of fresh air, thus causing a continuous change in the air of the drawers for the successful oxygenation of the blood. Underneath a perforated zinc bottom in each

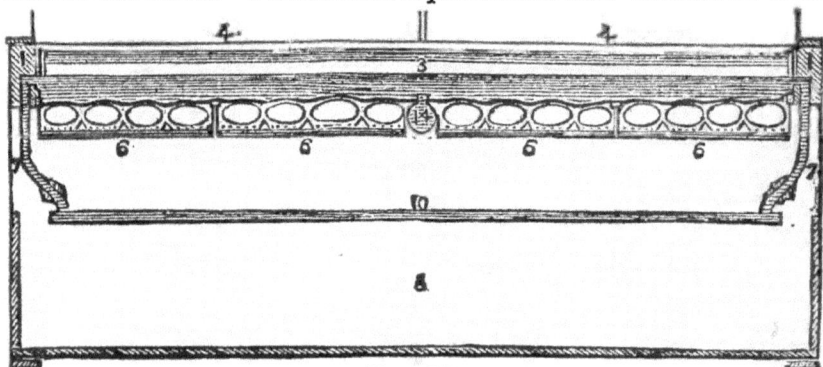

FIG. 78.

drawer is placed a layer of finely sifted garden soil, to the depth of about half an inch, to be constantly kept damp (not wet), and which diffuses a genial moisture to the air and eggs during the whole period of incubation. Each drawer is capable of holding between forty and fifty eggs of the smaller breeds of poultry, or about forty duck eggs; thus this Incubator can contain at one time from one hundred and fifty to two hundred eggs, and can hatch any size of eggs from a swan's to a sparrow's.

7. Four pipes to convey the warm water from the cistern above to the Artificial Mother below, which is seen in Figs. 76 and 78, No. 10.

8. Nursery for chickens, the front containing glass, and the back wire netting or wood (No. 9, Fig. 76). When more room is required for chickens, an extension equivalent to half the length of the Incubator can be attached to each end, thus doubling the area of the run.

10. Artificial Mother (sloping in the direction of the arrow from about eight inches high at A to three or four inches at B, see No.

10 on Fig. 76), heated by the water flowing from the tank above through four pipes (7).

11. Return pipe, conveying the water from the Artificial Mother to the boiler to be re-heated.

12. Boiler, containing three flattened elbow tubes, each one placed above a lamp wick.

13. Lamp for the burning of paraffin, petroleum, or other oils, containing three wicks, each two and three-quarter inches broad, the whole held in position by four rods fitted into corresponding tubes attached to the lamp, and retained by two sliding catches.

14. Inlet pipe, joined to the boiler by a brass coupling, conveying the heated water by five graduated openings or inlets to the cistern, whence it flows on both sides of the centre to the ends of the tank.

15. Heat regulator, a pill-box shaped brass vessel, having, as a lid, a flexible diaphragm, which acts by means of a rod upon the slide or cover of lamp No. 16, Fig. 76.

16. Graduated slide or cover for lamp. As the heat expands the air, or any other fluid that may be used, in the long tubes within the cistern, the only non-resisting medium upon which it can act is the flexible diaphragm on No. 15, causing it to protrude in the direction of the dotted line, and thus propelling the slide or cover (No. 16) inward upon the flames, and thereby cutting off the heat. As the heat of the water in the cistern or tank subsides, the atmospheric pressure, re-acting on the diaphragm, causes the slide to be withdrawn, and thus the size of the flames, and consequently the heat, is again increased.

17. Waste pipe to run off the water at the end of the hatching season.

18. Tray, made of the same material as the tank, surrounded by a wood frame in which to place the eggs after the birds have chipped the shells, and where the chickens can remain till they are dry, when they may be removed to the Mother below.

Where gas can be obtained and would be preferred, a regulator for it is made instead of the oil lamp.

We have no doubt but this is the best of the English machines, yet its complication of parts and great cost will prevent its ever becoming popular with those who have the most use for an Incubator.

THE SCOTIA INCUBATOR

is of the same nationality as the last. It is described in Brown's book on Artificial Incubation as being in appearance like a very flat box, with one drawer in front, and an arched hole at each end

of it. It is about three feet in length, half that in width, and
fifteen inches in height, the outer case being wood entirely. A
tank, about one and a half inches deep, is fitted in it, extending
the whole length of the machine, which is heated by lamps placed
in compartments of which the arched holes are the entrance.
These lamps are ordinary paraffin lamps, without chimneys.
The egg drawer is placed between the two lamp compartments,
and is sufficiently large to hold one hundred eggs. In the bottom
of this drawer sods are placed, or earth, and kept constantly
moist; above this a layer of straw is laid, on which the eggs are
put. Ventilation is given over the ends and back of this drawer.

FIG. 79.—CHRYSTY'S DYDRO-INCUBATOR.

The great fault in this machine is its defective ventilation and
lack of regulator.

CHRYSTY'S HYDRO-INCUBATOR

is a London, Eng., invention, of which Fig. 79 is an illustration.
In describing its parts, A is the exhaust pipe used in emptying the
cistern; B, a brass cock for drawing off water, previous to replenishing the cistern with boiling water; C, a glass gauge, with
a marked scale D at the side, to give the height of water in the
cistern; E, tube through which cistern is filled; F, vent tube for
escaping air; G, egg drawer; H, thermometer; I, air holes for
ventilation; J, flannel on which the eggs lie; K, stand or box on
which the Incubator sits; L, earth trays in bottom of drawer.

The mode of operating this machine is to fill the cistern with
boiling water, which raises the heat to about 120° to 130°; after

allowing it to cool down to 102°, keep the heat at about that point by drawing off every twelve hours a sufficient quantity of water, and replacing it with an equal amout of boiling water; the quantity requiring to be taken out varies with the outside temperature, for while a very small quantity suffices in summer, or if kept in a warm room in winter, if kept in a cold place it will take a large quantity to keep it going.

In the hands of careful operators, this machine has achieved some excellent results. Its defects are: the large quantity of water required to work it in cold weather, and the trouble of getting it hot twice a day; the insufficient moisture given off by the earth trays, and the position of the thermometer, which can only be seen by opening the drawer.

HOWELL'S GEM HATCHER

is very similar to the last described. The difference consists mainly in the egg drawer, which is a tin tray, the bottom covered with straw, on which the eggs are laid; the arrangement of moisture pans in the egg chamber is much superior to the earth of the former.

The inventor claims that, by reason of improvements in construction, much less water is required to work it than Christy's, which, if so, is a great advantage.

The machine holds about twenty-five gallons of water, the large body of which, doubtless, retains heat longer, and therein gains the advantage.

THE VOITELLIER INCUBATOR.

This is a French invention, and, like the last two described, is also a Hydro-Incubator.

It is described in Brown's work as "a strong deal chest, thirty-three inches square and twenty-one deep, containing a cylindrical zinc cistern, packed tightly round with sawdust; this hot water tank is a foot in depth, and has twenty inches inside diameter; it exactly fits, and rests upon a circular wooden frame four inches in height, and this is the "hatching nest", in which eggs are placed. It will hold about one hundred fowls' eggs, or seventy-five duck eggs.

"Two movable glazed frames, fitting one over the other, are provided at the top of the machine, allowing either of access to the eggs, when lifted off, or of a glance at the thermometer. The inlet pipe is at the right hand top corner, and the only outlet is at the bottom. In the centre of the front is a pipe to supply air, but the latter is always warmed, before being admitted into the Incubator, by having a considerable length of pipe running alongside of the cistern. The air has no direct contact, either, with the eggs, inasmuch as the pipe rises nine inches inside, and a current is secured at the top by the aid of a very small piece of piping, through which it is supposed the steam from the hot water escapes, and so dampens the eggs sufficiently.

The cistern holds twenty gallons of water. The nest is prepared by placing an inch of sand in the bottom, which is kept damp, over which is put straw or fine hay on which to lay the eggs ".

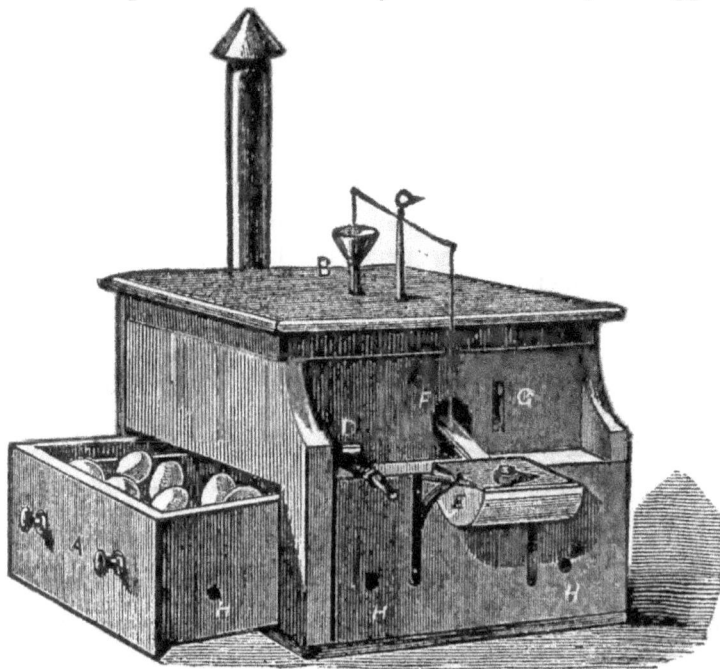

FIG. 80.—CASHMORE'S INCUBATOR.

The same remarks apply to this as to the two previously mentioned. It requires some experience and also "knack" in determining just how much boiling water is daily necessary to keep up the temperature, and especially in variable weather.

CASHMORE'S INCUBATOR,

illustrated at Fig. 80, is the last of the foreign inventions we shall notice.

The machine is two feet square and one and a half high. The egg drawer is shown in front, partly open; the bottom of this drawer is of perforated zinc or wire cloth, over which is laid a piece of flannel or felting, on which the eggs are placed.

A tray filled with damp earth is under the eggs to keep the bottom cool and moist. The tank is filled through the funnel and tube *B*, which is also used for a regulator. *C* is the pivot for a lever, connecting at one end with the float in the funnel and tube *B*, and at the other end by a piece of wire with the pipe of the lamp, on which is placed the burner. This pipe and burner is inserted into the tube *F*, which connects with the chimney *I*. *E* is a benzoline lamp, carefully balanced on the two brackets shown in the cut. The thermometer is shown at *G*, and the ventilating holes at *H*. Inside the tube *F*, a quarter inch beyond the burner and a little below it, is placed a piece of brass shaped like a widespread inverted Λ.

The mode of working the machine is as follows: The cistern is filled, through the funnel *B*, with boiling water, which is allowed to cool down until it is at the required temperature; the lamp is lit and placed in position, and the connecting wire from the lever is hooked on to the pipe or burner tube of the lamp. When the machine gets hotter than the temperature at which it is set, the water expanding raises the float, and, by the action of the lever, depresses the burner, which passes partly under the Λ shaped piece of brass, thus cutting off nearly all the flame. As the heat falls the water contracts, the float falls, and the burner comes back into its first position, and the flame attains its full size.

This is, probably, the most simple of the "self-regulating" machines. It has its defects, however, the principal ones being the inequality of heat in different parts of the egg drawer, and the evaporation of the water from the open tube, which creates the necessity for frequent replenishing of the water supply, and resetting of the regulating apparatus.

TRUESDALE'S INCUBATOR.

This machine is a California invention. Its shape is that of a bureau, three feet high and two feet square. There are three drawers, in which are movable trays for the eggs; these trays hold one hundred eggs each. Above each drawer are tanks of water, which are connected with a galvanized iron boiler, heated by a lamp. By a proper connection, the water is constantly in circulation between the boiler and tanks. There is no regulating attachment.

TATHAM'S INCUBATOR

is another Pennsylvania invention, its principal features being the division of the water tank by a horizontal partition of metal, perforated at each end, through which the water has to pass to reach the return flue; this flue passes down and under the egg drawer, into a moisture pan, which it traverses to and fro a number of times before passing out to be re-heated. The heating apparatus is the continuation of this same flue or tube, arranged in a spiral coil, and placed over a peculiarly formed gas burner.

This burner is fed through a double supply tube, in one arm of which is an elbow-like depression, opening at the bottom into one end of a U shaped glass tube (this tube is almost identical with that described in the Graves machine). The long part of this tube is

FIG. 81.—RENWICK'S INCUBATOR.

filled with alcohol and inserted into the egg chamber; the U shaped end is outside, and filled with mercury. As the heat in the machine increases, the alcohol expands, and forces the mercury up so that it rises above the elbow in the main gas tube, thus cutting off the flow through that, and forcing it to pass through the other and smaller tube, in which is a stop cock, thus reducing the flame to as small a jet as is wished.

Another noticeable point is the ventilation. The heat radiated from a metal tank over the eggs passes over each end of the drawers downward through a flue, and then turning upwards, is carried out through a ventilator at the top. The supply of fresh air is received from a valve in the bottom, from whence it passes over the water pan and upwards between the drawers, and thence off through the flue above mentioned.

The machine has some very good points about it, but of its working qualities we know nothing. It is only adapted for gas, and, therefore, of no benefit except to those who live in villages or cities.

RENWICK'S THERMOSTATIC INCUBATOR.

Is illustrated by Figures 81, 82, 83, and 84. Regarding its construction and working, Mr. R. communicates the following:

This incubator was invented by Mr. E. S. Renwick, of Millburn, N. J., the well known scientific expert in patent causes, and it is the subject of four patents, viz: No. 193616; No. 210559; No. 217-148; No. 224224. Its introduction marks a new departure in the art of Artificial Incubation, and, as it has been used successfully by the inventor for five years, its construction is worthy of careful study. Mr. Renwick's early experiments, led him to the conclusion that the generally accepted systems of ventilating Incubators, and of regulating the heat by varying the position of the lamp wick relative to the wick tube, or by moving the wick tube upon the wick, were not as reliable as they should be, and his Thermostatic Incubator is characterized by the following leading peculiarities; viz:

First, the ventilation of the incubating chamber by distributing the fresh air at its top, and drawing off the foul air at or near its bottom.

Second, the regulation of the heat of the Incubator by alternately permitting the heated gases from the lamp to pass through the incubating chamber, and to escape without passing through the chamber, but without altering the position of the wick or the wick tubes.

Third, the automatic supply of moisture at the top of the incubating chamber, where the air is distributed, and in exact proportion to the amount used; so that the air is charged with moisture before coming in contact with the eggs, and consequently the wetting of the eggs is dispensed with.

Fourth, the support of the eggs upon rollers connected with each other, so as to turn simultaneously in the same direction. Hence all the eggs in one tray or drawer, can be turned simultaneously by one revolution of a crank or key, and without the necessity of opening the Incubator.

The regulating mechanism for opening and closing the valves

which regulate the heat, contains various new features which are fully described in Mr. Renwick's patent, No. 210559; but a full description of them would occupy more space than we have to give to the subject. Those who wish information on the subject, can readily obtain it by procuring a copy of his patent.

The last pattern of Thermostatic Incubator used by Mr. Renwick, has the heat distributed through it entirely by the air employed for ventilation, and dispenses with the use of the water vessels generally employed to distribute the heat by the circulation

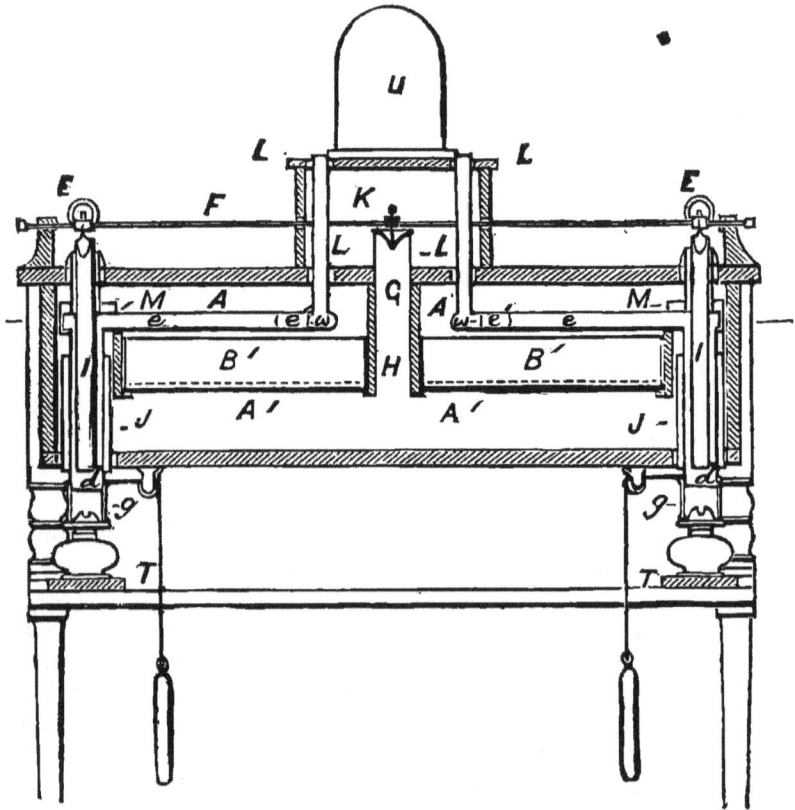

Fig. 82.

of hot water. It has an incubating chamber divided by four upright partitions, which form the slides for the four drawers B, B, in which the eggs are placed. If the mechanical system of turning the eggs is used, three of the drawers are traversed by series of rollers about an inch in diameter, and about two inches from centre to centre; and the eggs are laid in rows upon these rollers. If the rollers are not used, the bottoms of the drawers are formed of wire gauze, or of perforated zinc, so that the air may readily circulate downward through the egg drawers. The central two partitions between the drawers, are separated by a space H, in which the thermostat, which controls the temperature, is placed. This space communicates by a ventilating nozzle G, fitted with a valve K, with the casing of a chicken drawer, arranged on the top of the Incubator; and the top of this casing has ventilating holes in it for the escape of the foul air.

The Incubator is heated by means of two kerosene lamps T, T, supported upon a counterbalance lamp gallery, which is suspended under the Incubator, so that either lamp can be readily removed and replaced by depressing the gallery and allowing it to be raised by the counterbalance weight. Each lamp is provided with a short chimney g, which is applied to the lower end of an upright flue d. The upright flue connects at its upper end with a horizontal T shaped flue, e, which extends through the top of the incubating chamber, and has an escape pipe, I, from which the spent products of combustion escape. Each upright lamp flue d is surrounded by an air flue J, which is open at the bottom so as to receive the external air; it is also opened at the top so that the air which draws through it escapes at its upper end into the upper part of the incubating chamber. The air flue, and the lamp flue within it, form a small hot air furnace, so that the air which thus enters the Incubator is warmed in its ascent. The entering air distributes itself in the top of the incubating chamber, forcing the air therein downward through the egg drawers and into the bottom of the central space which forms a chimney, by which the air escapes. Hence, when the Incubator is in operation, there is a constant circulation of air from the upper part of the Incubating chamber downward, against the natural tendency of hot air to rise. This downward circulation gives time for the particles of air of equal temperature to arrange themselves in strata of practically equal temperature horizontally, but progressively cooler as they approach the bottom of the Incubator. Hence, the eggs are heated hottest at their upper sides, while the temperatures at different parts of Incubator of the same level, however widely separated, rarely differ more than a degree; and are generally within half a degree of each other.

In order that the heat may be regulated, each upright heating flue, d, is traversed centrally by a waste heat chimney, I, whose upper end is controlled by a valve, E. When these valves are dropped the waste heat chimney is closed, and the hot gases from the lamps are compelled to traverse the heating flues and to heat the interior of the Incubator. When, however, the valves are raised, the waste heat chimneys are opened, and the hot gases take the direct course through the chimneys and escape without heating the Incubator. Hence, the opening and closing of the valves of the waste heat chimneys determine the heat; and these operations are effected by the valve mechanism arranged under a glass shade, U, on the top of the Incubator. The opening of the waste heat chimney is attended with another useful effect; thus, when it is open, the draught is stronger than when it is closed; and in the former case the current of cold air in the cone or deflector of the lamp infringes upon the flame with greater force and reduces the volume of the flame; thereby producing the same effect upon the flame as the lowering of the lamp wick; while the reduction of the force of the draft incident to the closing of the valve, permits the flame to increase in volume. The

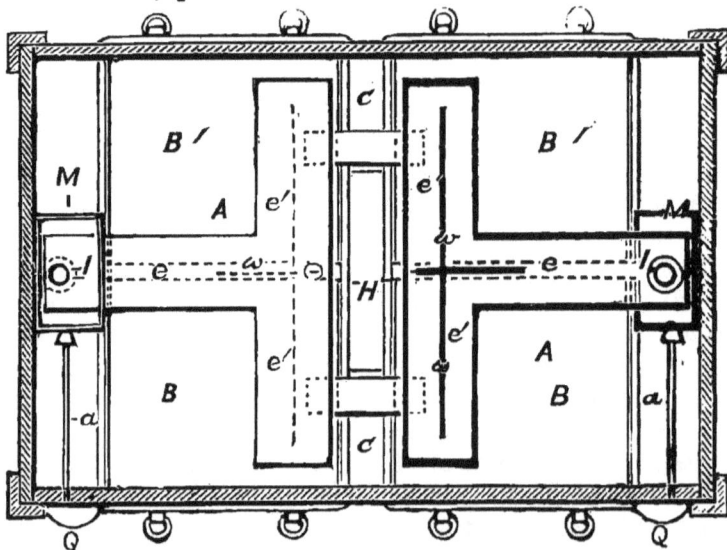

FIG. 83.

valve K of the ventilating chimney nozzle G is made smaller than that nozzle, so that it can not wholly prevent the escape of air, and consequently the ventilation can never be stopped. This valve is connected with the same shaft that operates the waste heat valves E, so that when they are opened the ventilating valve is also opened, and the ventilation is increased; and when they are closed the ventilating valve partially obstructs the chimney nozzle, and reduces the quantity of air escaping from Incubator; thus facilitating the rise of temperature.

The valve mechanism is operated by a weight, and the times when this weight is permitted to open or to close the valves of the waste heat chimneys, and of the ventilating chimney, are determined by a compound thermostat composed of bars of brass and vulcanite. Two compound bars arranged back to back are used;

each bar being composed of a strip of brass and a strip of vulcanite, 24 inches long and 1 inch broad, rivetted together at intervals of an inch. These compound bars are supported at both ends, and are connected at their cenrtres with a series of two multiplying levers, the fulcrum of the first of which is connected with one compound bar, while the shorter arm of the same lever is connected with the other compound bar; hence, the lever is moved simultaneously in opposite directions by the greater or less curvature of the two compound bars; and the extent of movement is about double as much as it would be with one bar. A counterpoise for the weights of the thermostatic bars is employed, so that they are relieved of the strain of moving their own weights, and are consequently very sensitive to changes of temperature. The end of the second multiplying lever is connected by a light rod, with a detent which releases the valve mechanism whenever the valves are to be opened or closed. Mr. Renwick uses a weight heavy enough to open the valves with absolute certainty, and he found that the friction incident to the strain of this weight, affected the sensitiveness of the mechanism, hence he interposed a detent shaft (turned by a light spring,) between the main valve shaft and the detent of the thermostat, so that the detent is relieved of the strain of the operating weight. A light shifting weight also is employed, to counteract the effect of the slight residual friction, by helping the thermostat to move the detent in alternately opposite directions, as the heat rises and falls. To prevent the valve mechanism from moving so rapidly as to create a jar, a speed regulator is employed, consisting of a small paddle wheel, running in a semicircular trough containing glycerine. This contrivance effectually prevents excess of speed, while it is practically frictionless.

The supply of moisture to the Incubator is furnished by two glass fonts $N\,N$ set upon the top of the Incubator. Each font has a discharge pipe h, whose point dips into a small basin Q, which is connected by a tube a with an open topped pan M, surrounding the head of the upright lamp flue. The water stands at the same level in the internal pan M, and in the external basin Q; and as the water in the former evaporates, a supply runs in from the latter. Whenever the water sinks in the external basin sufficiently to unseal the end of the dip pipe h, a little air enters through this pipe into the font, and a corresponding quantity of water descends to keep the water in the basin and in internal pan at the proper level. The Incubator consumes nearly four quarts of water in 24 hours. The fonts are filled once a day, by closing the point of the dip pipe with the finger, withdrawing the cork of the font, pouring in the water, and replacing the cork; after which the finger is withdrawn to open the dip pipe. The greater the heat, the greater the quantity of water which is evaporated; so that the supply from the fonts is varied automatically. Mr. Renwick in his first Incubator, used stop-cocks adjusted by hand to regulate the supply of water directly to pans, as described in his patent No. 193616; but the plan above described operates perfectly, and obviates the necessity of any other operation than replenishing the fonts once a day.

The temperature employed by Mr. Renwick varies continually from about 98° to 106°, these extreme temperatures occurring from half hour to two hours apart. His mechanism can be adjusted to run within an extreme variation of three degrees, as indicated by a sensitive thermometer, but he believes that a greater variation of temperature is more in accordance with the variations that occur when eggs are hatched under a hen, by the change of position of the eggs from the centre to the rim of the nest. He finds that the temperature inside the egg does not vary more than two degrees with a variation of eight or ten degrees in the surrounding air, provided the extreme temperatures are not of too long duration; and, in the practical use of his Incubator, the variation in the temperature of

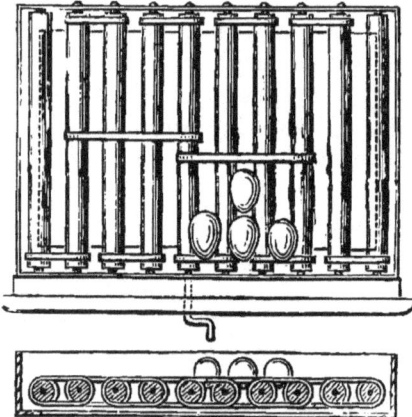

FIG. 84.

the eggs themselves does not exceed one degree. He also finds that, with the ample ventilation and supply of moisture maintained in the incubating chamber of his apparatus, an occasional increase of temperature to 110° is not fatal to the eggs.

Mr. Renwick's system of turning the eggs is represented in Fig. 84, the eggs being supported on rollers, which are connected by endless bands of tape or of India-rubber; the pivot of one of the rollers is extended through the front of the egg drawer, and is fitted with either a crank or with a clock key, by turning which all the eggs in the drawer are turned simultaneously. This system of turning by rollers was patented to Mr. Renwick in Patent No. 224224. The eggs are left on the rollers for nineteen days, when they are by preference transferred to a drawer having a perforated bottom, upon which the chickens are supported with ease to themselves, and with the capacity of moving about, when they emerge from the egg shell.

After the chickens are hatched, Mr. Renwick brings them up in what he calls a " Ventilating Brooder", in which the chickens are kept warm by hot air heated by a small lamp. The air is introduced through a perforated floor, on which the chickens are supported, and they are covered by an inclined board whose under side is lined with plaited carpet. The covered brood chamber communicates at one side with an enclosed run in which the chickens are fed. This " Brooder" is fully described in Patent No. 215,070.

SMITH'S INCUBATOR.

This is one of the latest inventions, and was exhibited the past winter for the first time. In general shape it resembles the Eclipse. It consists of a case enclosing two tanks of water, an upper and a lower one; the first to give heat by radiation to the top of the eggs; the second to vaporize water and keep the air charged with moisture. Each tank has a capacity of about fifteen gallons; the large amount of water being used to prevent sudden changes of temperature in the egg drawer, which is between the two tanks. These tanks are both open at top, so that the evaporation of water is constant and copious, and renders necessary frequent replenishing.

The regulation is effected by electricity. A "pyrometer" (another name for a thermostatic-bar) is used as a circuit-closer, and the electric current acting on a magnet, opens and closes the ventilator.

Over the top of the machine is an enclosed space which is used as a temporary artificial mother. This is warmed by the waste heat from the machine.

The cost of running a machine of two hundred egg capacity is said to be about eight or ten cents per day.

BATCHELLER'S PERFECTION INCUBATOR.

Of this machine we know nothing, save that it is offered for sale by the inventor in one of the Western poultry papers. We have been unable to get a description of it from either the inventor or any one else, hence conclude that it will not bear investigation.

THE FAVORITE INCUBATOR, (See Fig. 85),
is a Yankee adaptation of ideas gleaned from other machines.
The Boiler and lamp are taken from the Centennial; the tank
and egg-tray from the Novelty or Carbonnier's; and the regulator is an adaptation of Guest's patent fire alarm or heat-indica-
tor. The following description is taken from the circular of the
manufacturers.

"The heat in the machine is applied from the top, with perfect uniformity throughout the egg-drawer, no greater in one
part than in another, and under the control of an automatic
regulator.

The moisture is supplied from below the eggs, in just sufficient quantities, and at the proper degree.

The ventilation is steady and unchangeable, a constant current of air passing through the egg-chamber at all times.

The case of the machine is of wood; an inner one of pine, and an outer one of black-walnut.

Fig. 85.—THE FAVORITE INCUBATOR.

The regulator governing the heat in this Incubator, is composed of a group of bars in the top of the egg-chamber, and made
of a material that is very sensitive to heat and cold; its action is
positive, opening and closing the ventilator, and graduating the
flame of the lamp, thus checking the advance or decline of the
temperature. The mechanical part is regulated by a thumb
screw on the outside of the machine, which allows the temperature in the egg-chamber to be fixed at any desired point. When
properly adjusted, the expansion bars affected by the heat, act
upon an escape lever, releasing an arm, which passes from one
side of the lever to the other, at the same time turning down
the lamp flame, and opening the ventilator, allowing the hot air
to escape from the egg-chamber. It remains in this condition
until the heat has fallen one to three degrees, when a reverse action of the escape occurs, causing the arm to return to its former

position, the ventilator to close, and the flame of the lamp to be turned up; this movement takes place every fifteen to thirty minutes, and goes on continually, by means of power transmitted by a simple reel and weight."

The machine is a very good one, but being a plain infringement on several other patents, it has never been patented, and its use may at some time bring trouble from that cause.

THE SUFFOLK INCUBATOR (See Fig. 86.)

finds its birth-place on Long Island, New York State. It is described by the inventor as follows:—

"The Incubator is strongly made of yellow pine and walnut. There is absolutely nothing to get out of order and give trouble in the whole incubator; can be managed by any person of ordinary intelligence; it will hatch all the eggs that would hatch under the most favorable circumstances in the natural way.

The incubator is heated by hot water in galvanized iron tanks, with perfect uniformity of heat through the egg drawers.

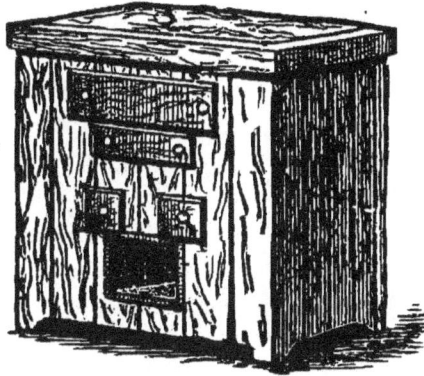
Fig. 86.-THE SUFFOLK INCUBATOR

Two drawers are arranged at the side of the lower section of the tank and receive the young chickens directly after they are hatched, the chickens being dried in these drawers.

The moisture, which is a continuous evaporation supplied by an earth drawer from below the eggs, and at the proper degree of heat, so that the eggs do not require sprinkling at any time.

The ventilation is steady, as a constant current of air is passing through the egg drawers at all times, the air passing in at the bottom and around the lower tanks and over the moisture pan, keeps the air at an even temperature, before it passes through the egg drawer and out at the top of the incubator through the air chamber, giving a constant current of air without chilling the eggs."

The last paragraph of this description is rather vague; but we give it as we find it. The machine has no regulator and like others of its class requires more or less watching.

### THE WHITE MOUNTAIN INCUBATOR		(See Fig. 87.)

is another adaptation of other people's ideas. In all the essential points it is a minature "Eclipse" differing simple in shape and and arrangement of egg-trays, etc. It is regulated by a battery and electric circuit. As its name indicates, it was "hatched" out in the "Old Granite State."

### THE ACME INCUBATOR		(Shown by Fig. 88.)

is the invention of the writer. It is a hot-air machine, and was designed to meet a call for a cheaper machine than the Old Centennial.

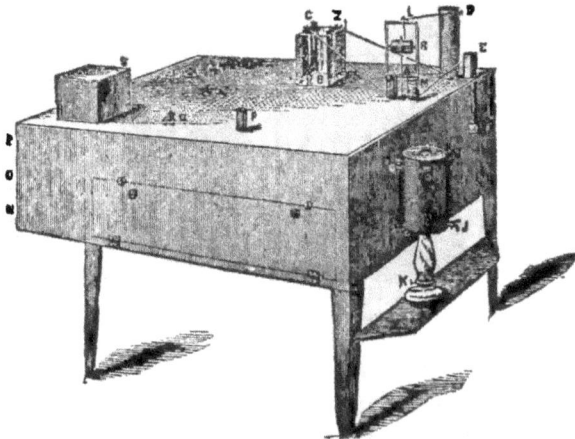

Fig. 87.—THE WHITE MOUNTAIN INCUBATOR.

Fig. 88 shows a one-hundred and fifty egg machine with doors closed.

Fig. 89 shows the inside of egg-chamber which is explained as follows:—

The heat is generated by the lamp L, which has an ordinary "B" burner (one inch wick). A copper drum is heated by this lamp, from which the warm air radiates and passes upwards through the hot-air chamber R, where the evaporating trough H, divides the current, and charges it with moisture. The shield U, deflects the rising air over the drawers D D, in which the eggs are placed. By a peculiar arrangement, the air is caused to pass out of the sides of the egg chamber at Y Y, and thence through the ventilators V V.

The smoke and gas from the lamp are carried off outside the Incubator and cannot by any possibility enter the egg chamber. Here has been a great scource of failure in machines of this class; the fumes from the lamp entering the egg chamber and killing the chicks.

The hot-air chamber is closed by the small door E; the inner door B, is then closed and fastened, and then the outer door A, which is double and packed. By this construction the eggs in the first row next the door are equally warm as those in the back of the machine.

The small windows G G, give light enough to see the thermometers, and also to examine the eggs when hatching, without opening the inner door, and thus cooling off the egg chamber.

The lamp L is attached to and suspended under the heating box by spring catches, one of which is shown at X, and can be detached or put in place in a few seconds. It can be filled without being taken off. In packing, the lamp is detached, the heating box also slides out, the legs are taken out of their sockets, and all are packed inside the machine; thus reducing the size of the box, to about half that of any other incubator of the same capacity. As before stated, the Acme is constructed entirely of metal; is double cased, with a space of three inches non-conducting packing between the cases.

Fig. 88.

THE ACME INCUBATOR.

It is fitted with the same kind of thermostatic or regulating-bar which has proven so successful in the Centennial. The regulating apparatus is also similar, but much more simple.

The egg-drawers are an entirely new invention of my own, and can be used throughout the entire time of incubation. With the egg-turning-trays heretofore offered, it was necessary to substitute a plain drawer or tray with a tight bottom when the chicks were hatching. Another advantage of the Acme turning-trays is that they will hold one more row of eggs than those of any other make. In an incubator, the size of the No. 1 Acme, this would make a difference of fully twenty eggs. The entire con-

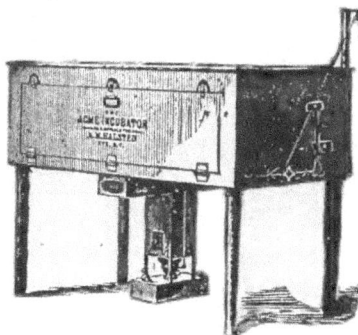

tents of both drawers in the Acme—150 eggs—can be turned in ten seconds. A broad patent was granted on this machine May 23d, 1882.

HAIGHT'S PATENT INCUBATOR

is the invention of Henry J. Haight, of Goshen, N.Y. We have no full description of it, Mr. Haight, not yet having placed the machine on the market; although rather complicated it is said to achieve good results. It is regulated by a thermostat.

The egg-tray is worthy of particular mention; it is a rack

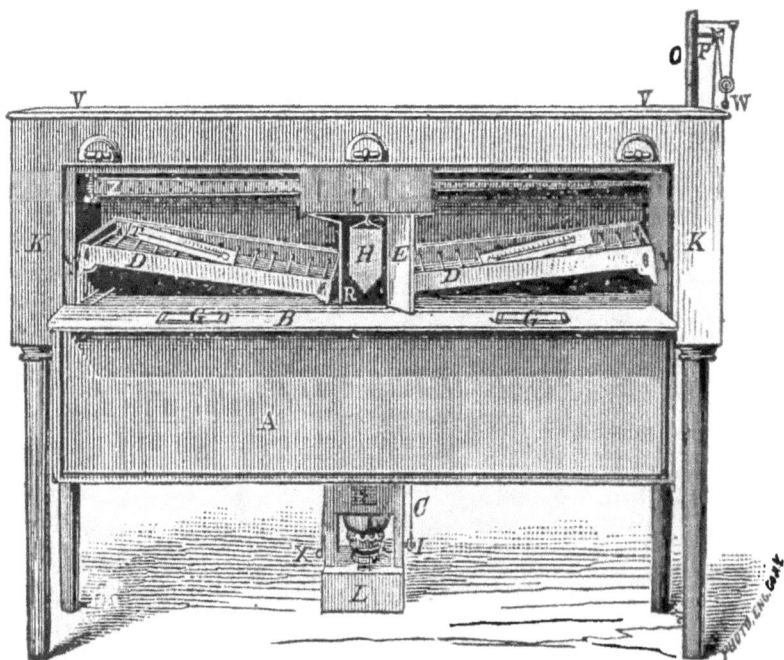

Fig. 89.—THE ACME INCUBATOR. (Inside View).

covered with coarse muslin or sacking on which the eggs lie; this rack is suspended by a bar or axle across the centre, the ends working in journals; one end of the rack being held in place by a spring bolt. A duplicate rack of same construction is laid over the first and secured to it, thus placing the eggs between the two trays; the bolt is then drawn and the two trays revolve, or turn half of a revolution, thus turning the eggs upside down, and leaving them resting on the sacking of the second tray: the first tray is then unfastened and taken off, and the eggs put back in the machine.

The machine and turning-rack are both patented.

J. M. HALSTED'S NEW SELF-REGULATING INCUBATOR
(See Fig. 90),

is a California product, hailing from Oakland, and was patented Aug. 8th, 1882.

The inventor says of it:—

The Incubator is a hot-air Machine, warmed by a kerosene lamp, and burns about 1½ gallons of oil to hatch 100 eggs, or 3 gallons to 250 eggs, in this climate.

It is made of five of the best non-conducting substances and constructed so thoroughly that years of constant service will not impair its efficiency. The front is furnished with double glass doors through which the eggs and thermometer are visible without opening the machine. By the scientific manner in which the heating apparatus is constructed, every particle of heat is utilized and a great saving of oil is effected. The air thus warmed is automatically moistened by an ingenious device, before entering the egg chamber, through which it passes in a constant current over the eggs and then through the ventilators, which are always open, yet placed in such a position that no cold air can enter. The moisture can be increased or decreased as desired, which being done automatically, avoids the necessity of sprinkling the eggs daily by hand.

Fig. 90.

J. M. HALSTED'S NEW SELF-REGULATING INCUBATOR.

The lamp gives sufficient heat to use the machine successfully in the coldest climate, and yet is so constructed that it works equally well in the warmest.

The formation of the heating apparatus is such that neither smoke nor gas can enter the egg chamber, in which the air is constantly changing, therefore it must always be pure and wholesome. Underneath the eggs a current of cool—not cold—air is kept circulating, which, as it becomes impregnated by carbonic acid gas from the eggs, passes out through the bottom ventilators.

The new Patent Regulator is the Perfection of Simplicity, is strong, reliable and will last a lifetime; it is connected directly with the lamp and turns the flame up or down with the least variation of temperature.

Above the egg chamber is an artificial brooder, in which the young chicks can be placed as soon as dry.

LA BARGE'S INCUBATOR

comes to us from Missouri. It was patented Dec. 21st, 1880. It is a rather complicated affair, regulated by electricity.

Its chief claim to notice is in its egg-turning arrangements. The eggs are placed in what the inventor calls "hammocks" which are suspended from frames or "cradles." These are in tiers, and are hinged at one end, and the other connected with levers, which, elevating the free end causes the eggs to roll in one direction, and lowering it, they roll the reverse way. The movement is caused by a clock, set to act every twelve hours.

THE PACIFIC INCUBATOR   (Fig. 91.)

as its name indicates, originated on the Pacific Coast in California.

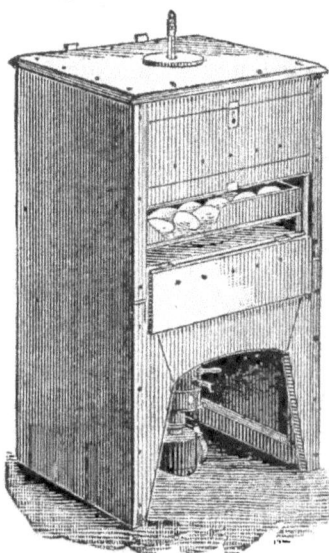

Fig. 91.
THE PACIFIC INCUBATOR.

"The boiler and water chambers are constructed of heavy galvanized iron, and the heat generated by the hot water circulation is applied to the eggs from above, thus imitating nature as far as possible, and the temperature is ascertained by inserting the thermometer in the water chamber at the top of the machine, as shown in the accompanying cut.

The exterior case is made of Sugar Pine, and finished so as to resemble a cabinet or chest of drawers, and with reasonable handling will last for twenty years.

The heat is applied from one kerosene lamp, situated under the center of the boiler in the two smaller sizes, and from two lamps (one on either side) of the largest sized Incubator.

The temperature is regulated by turning the wicks of the lamps up or down as may be necessary; but this requires very little attention, and the machine can be left to itself for hours together, and there is no necessity for touching it during the night, a slight turning down of the wicks in the evening being all the care that is required, as a fall in the temperature of two or three degrees during the night does no injury."

As will be seen from the foregoing description, this machine has no regulator. While it might run with comparatively uniform temperature in the climate of California, in the country east of the Rocky Mountains, it would be absolutely impossible to control the temperature within ten or more degrees, unless it was constantly watched. We understand the machine has accomplished good work in the hands of its inventor. It was patented January 20th, 1883.

THE PARREY INCUBATOR (Fig 92.)

hails from Michigan. It is a Hot-Air Machine. The following description is taken from the circular of the manufacturer :

"The oven or egg-chamber is a wooden frame covered with the heaviest straw-board, both inside and outside, with a space between of an inch and a half, which is left on the sides simply as a dead-air space, but on top is filled with non-conducting material.

The doors, of which there are two, close against india rubber jams, thus making air-tight joint. In one of the doors is a window, through which the thermometer may be seen without opening the oven. Beneath the oven is the heat-er, a compound cylinder of iron, in fact three cyl-

Fig. 92.
THE PARRY INCUBATOR.

inders, so arranged that though no smoke or fumes from the lamp can get to the eggs, there is a current of hot, pure air, continually ascending into the oven, and in so ascending it has to pass through a zinc tube, from which it is discharged near the top of the oven, from whence it is, by means of a radiator above and (in the large ovens) a distributer lower down, brought down to the eggs in such a way as to give a very uniform heat in all parts of the oven.

The lamp is below the heater, and for the smaller sizes of machines, is a lamp of tin, made expressly for the purpose. For the larger machines I use an oil stove of the most approved style.

In the zinc tube above mentioned, is a valve or damper, so hung, that as the regulator expands with the increase of heat, the valve drops shut by its own weight, and is again opened by the contraction of the regulator as the heat decreases.

The regulator (not shown in the cut) is a double bar which expands and contracts as the heat rises or falls, and operates a valve as described above. The bar is so arranged that by simply turning a thumbscrew, the valve can be made to open or close at any desired point."

This machine is not patented.

THE NONPAREIL HYDRO INCUBATOR (Fig. 93.)

which has its origin in Wisconsin, belongs to the class represented on pages 100 and 101, viz:—"Chrysty's," "Howells," and "The Voitellier."

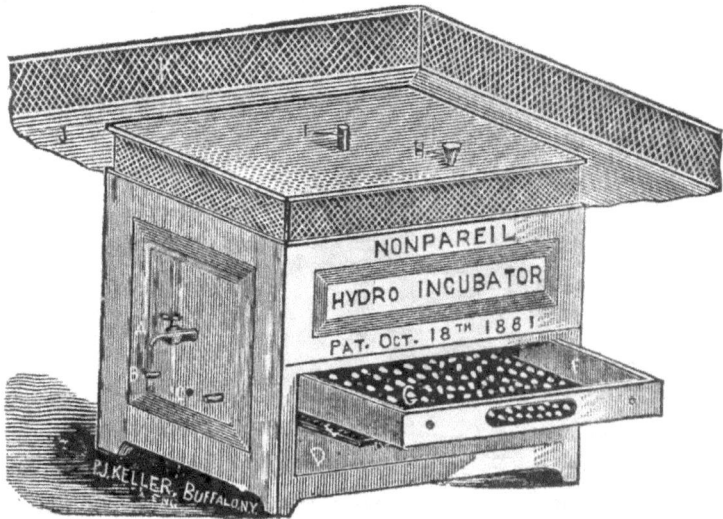

Fig. 93.

It consists of a large tank, packed down in some non-conducting substance; the tank being partially emptied, and re-filled two or more times daily with hot water, to keep up the heat, no lamp or regulator is used. It was patented Oct. 18th, 1881, and described in circular as follows:—

The form of the "Nonpareil" Hydro Incubator is seen in the cut, and is furnished with a large galvanized iron tank. The tank is six inches smaller each way than the outside frame, which space is packed with non-conducting material, as is also a three inch space on the top of the tank, which retains the heat in the tank a long time. The egg drawer, G, is below the tank, and is

one inch smaller each than the tank. The bottom of the drawer
is made of perforated zinc, this is covered with flannel, on which
the eggs rest. Under the eggs are the trays of water, E, with
cool air circulating over them, which evaporates the water, causing a moisture to arise which keeps the eggs moist. B B shows
the air tubes which passes through the side of the machine
through the packing and into the centre of the egg drawer, over
the eggs. Over the tank is the brooder, where the chickens
should be placed as soon as they are hatched. The brooder is
kept sufficiently warm by its position over the tank, thus making one machine do the work usually done by two separate machines, thus reducing the expense of raising the chicks. A, is

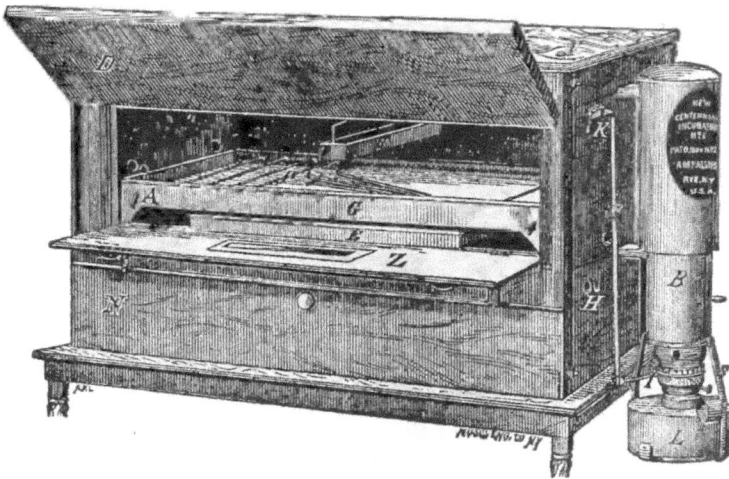

Fig. 94.—THE NEW CENTENNIAL INCUBATOR.
(Patented Nov. 14, 1882).

A—Thumb piece, or crank, on roller of turn-
    ing-tray,
B—Boiler.
C—Connecting rod from outside lever of
    rock-shaft to lamp.
D—Outer door.
E—Moisture pan under egg tray.
G—Egg tray.
H—Ventilator to nursery.
I—Lamp lever.

K—Outside lever of rock-shaft.
L—Lamp.
N—Door of nursery.
P—Rods or wires of adjustable egg tray.
S—Springs that hold lamp in place,
T—Moisture trough.
V—Ventilator flues from egg chamber.
Z—Inner door with glass window, (shown
    open).

the faucet for drawing off a portion of the water; D, false front
that closes up after the egg drawer and water trays are pushed
into their place; I, large air tube passing down into the egg
chamber; J, shelving for the chicks to run out on; K, lattice work

to keep chicks on the shelves; L, flexible curtains to allow chicks to run out and in under the brooder.

THE NEW CENTENNIAL INCUBATOR   (Fig. 94.)

is the invention of the author of this book.   It is the culmination of nearly twenty years' study and experiment.   Point by point has been worked out, tested and adapted, until the present machine is offered to the public, with the firm belief that there is nothing better to be had in the market either at home or abroad.

The reader will please observe that I do not claim—as do some of my modest competitors—to have the only incubator in the

Fig. 95.—THE EUREKA INCUBATOR.
(See Page 96).

market that will hatch a reasonable percentage of the eggs. Neither do I claim to have a perfect Incubator—as do others— for there is no work of man's hands that ever was or can be perfect; and he who claims his invention or creation as such, is too well satisfied with himself to see wherein he has made mistakes, and too obtuse to correct them or improve his work.

The inventor who considers his creation as perfect, and then rest upon his oars, soon strands upon the shoals of ignorance and self-conceit; while his wide awake rivals are carried onward by the tide of improvement, to success and fortune. Recognizing the above facts, I have made it my aim to give as perfect a machine as possible, to improve wherever I saw a necessity or an opportunity ; to add every convenience essential to success, and to turn out a thoroughly good and practical Incubator, at the lowest price consistent with good workmanship.

THE NEW CENTENNIAL INCUBATOR

is made in two parts ; an inner case of galvanized sheet iron, covered by an outer casing of wood ; with a dead-air space between the two cases. It has double doors—an inner and outer one— the inner one being provided with a glass window through which to examine the thermometer and the eggs. The outside case is held together by screws, and long bolts bind the whole machine firmly together, so that if a leak should occur at any time, it can easily and quickly be taken apart, the necessary repairs made, and the machine put together again without bruising or defacing the case.

There no electricity, no clock work, no weights, pulleys, or double levers. A simple rock-shaft passes through the side of the machine, with a lever on each end of the shaft; one of which is connected with the Regulator, and the other with the lamp.

A thumb-screw in the back of the machine, on the outside, adjusts the regulator to any required degree of heat.

The Regulator is a combination of Thermostatic bars, so pivoted and linked together, that they multiply power and motion. In other words, in the No. 1 machine, I use four bars, each two feet long; but my arrangement of them makes them equivalent in movement and strength to a bar sixteen feet long. I thus get a motion and power sufficient to act directly upon the lamp and ventilators through a single lever, and do away with all clock work, weights, reels, pulleys, etc. This regulator is placed in position before the machine leaves the factory, and all the purchaser has to do is to connect it with the lever and adjust it. It is placed above the eggs, out of the reach of the young chicks. It is sensitive to the least change of heat, and instead of changing the flame from one extreme to the other—either very high or very low—as is the case in other machines, it regulates the lamp to give the required heat. The action is regular and graduated to the needs of the machine: if in a very warm room, a low flame is produced: if the room grows colder the flame increases; and if the temperature of the room continues to fall, the flame is increased until the full power of the lamp is turned on.

Ventilation is provided for, by taking in a current of pure air, which passing close to the tank, is heated before it comes in contact with the eggs. It is then drawn to the four corners of the egg-chamber and thence carried by tubes outside of the machine. By this device the sides of the egg-chamber receives the same amount of heat as the centre, and there are no cold-corners. The ventilation is constant, not fitful, and the air is always pure and sweet. The method of ventilating solely through an opening in the top of the egg-chamber has been modified. In using that system exclusively, the egg-chamber is alternately overheated by turning on the full power of the lamp, and then cooled down by opening the valve and allowing the hot air to pass off; thus making an unnecessary waste of fuel, as well as a constantly changing temperature.

Moisture is provided, first by an evaporating pan so placed that it receives a gentle heat from the return flue, and thus supplies a moderate amount of vapor constantly under the eggs, and second by a trough suspended above the eggs, where it receives the direct heat of the tank.

The turning-trays which were patented in the "Acme" Incubator, have been adapted to this machine. It is a very simple device: a band of any suitable material is laid upon the bottom of the egg-tray, and attached to, or passed over rollers at either end of the tray: rods are placed across the tray at proper distance apart to allow the eggs to lie between them: by drawing this web or band in either direction the eggs are rolled or turned. The entire contents of the tray can thus be turned in from three to five seconds, and without taking them from the machine.

These rods are movable and can be placed at any desired distance apart, thus adjusting the spaces to suit eggs of any and all sizes.

The egg trays are all on one tier; not one above the other. It is a well known scientific fact that heat always rises, and that therefore it is impossible to keep a room or chamber at the same temperature at varying elevation. Hence a box or chamber heated to 103° at a certain height, will vary three to eight degrees at six inches above or below.

The space under the egg-chamber has been utilized as a nursery or temporary brooder for the newly hatched chicks, in which they may be kept for a week, or longer, if desired. This is heated by the flue which carries the hot water from the tank back to the boiler, and supplies the heat to the chickens' backs as designed by nature—not under their feet, as is the case with many machines, making the chicks weak and sickly.

Another new feature is the arrangement of the heating apparatus, so as to use the same lamp in either winter or summer. A movable cap over the boiler, when removed, creates a draft of cooler air which carries off the surplus heat: when in place it reverses the draft of air, causing a heated current, and doubling the heating surface of the boiler. This arrangement is applied more particularly to the No. 1, or one hundred egg size. The larger sizes, No. 2, three hundred eggs, and No. 3, five hundred eggs, are fitted with several burners so arranged that only one or more may be used as needed. Both the No. 2 and No. 3 machines have been run in a cool room with only one " D " (one-and-a-half inch) burner, showing a very small consumption of oil.

These several features are all secured by patent.

For further description and prices, address the inventor, A. M. Halsted, Rye, New York.

CHAPTER XI.

Houses, Yards, Location, etc.

It is not every location that is suitable for poultry raising. A low marshy place should be avoided; it will be almost certain to cause disease in the flock of old fowls, and make it nearly impossible to raise a paying percentage of the young. Place the buildings on an elevation, if possible; the grounds dry, and preferably with a southern or south-eastern slope. If the yards are bordered by water—a pond or stream, so much the better.

A rough piece of land well sprinkled with underbrush and rocks makes an excellent poultry yard, provided the rocks are not broken and piled, so as to afford a harbor for rats, weasels or other vermin. Small trees or bushes are desirable for shade and shelter. I advise planting currant bushes—also low growing evergreens. In my own yards I have peaches, plums, pears, and quinces, and find the fowls are beneficial to the trees and fruit; I have good crops every year.

On new ground for immediate shade, I would plant quick growing vines—gourd, morning glory, etc., and train them on brush stuck into the ground so as to form a sort of low arbor.

In the yards for young chickens, set out tomato plants (or plant the seeds) and protect them until the vines begin to set their fruit; train them the same as advised above, and you will have a good shade for the chicks, as well as a bountiful and wholesome supply of vegetable food for them during the fall.

Regarding the size of the yards, we have to be governed by our limits: in other words, "cut our garment according to our cloth." If the land is available I would advise abundance of room; if not, do the best we can. I do not say it is impossible to raise a great number of chickens on a small piece of land, but the smaller the yard or yards, the greater the care, and the more work required necessary for success.

I would not recommend less than one-eighth of an acre for fifty hens and four cocks; I would prefer one-quarter acre for that number—yet I can (and have) done well with that number on a yard of less than one-sixteenth of an acre. I consider it possible to raise to the age of broilers, and market from a quarter acre of ground, fifteen hundred to two thousand chickens per year; but to do this, requires unceasing care and watchfulness, and the most thorough sanitary precautions.

To the beginner I would say, do not try to keep over fifty laying hens to a quarter acre of ground; nor to raise over five hundred chickens per season on the same limits, To do even this, the runs for the young stock should be divided into sections— about four will answer, and one of these sections spaded or

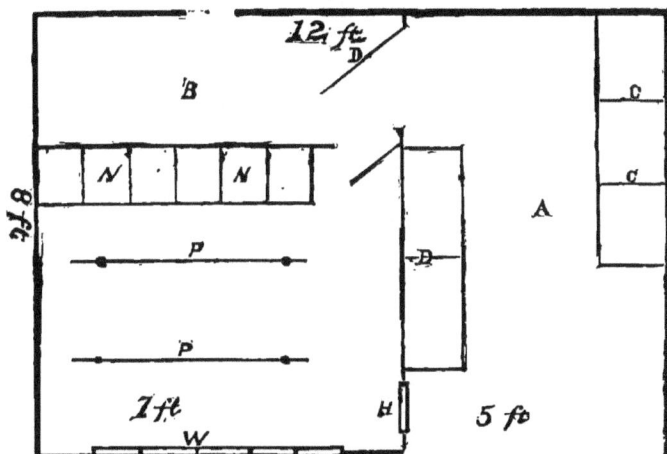

Fig. 96.

plowed, at least as often as once a week; and semi-weekly would be better. Sow this plowed ground with wheat, oats, barley or buckwheat, harrow or rake it in, and let the chicks scratch for it. One quarter to one-third of the run should be kept in sod— grass—unless you have a meadow or grass plot to turn them on pert of each day. The portable fence, illustrated by fig. 101, page 132, will be found very convenient in this work.

In the buildings to shelter the fowls, I would advise a number of small houses, rather than one of large size, for the breeding stock. A convenient as well as economical way, would be to build each house *double;* that is, to shelter two yards of fowls, letting the dividing fence join the house in the centre. Houses twelve feet long, by six feet wide, will make two apartments, each large enough to accommodate fifty hens and four cocks,—which are as many as had best be kept together.

This, however, is not always feasible, and it is more convenient to put each yard by itself. In such a case I would recommend a building planned like Fig. 96. The roof of this may be either double-pitch or lean-to, and the outside construction to suit the taste and means of the builder. The house is eight by twelve feet, divided into an open shed A, five feet wide, on one side of which is a dust bath D, and on the other side coops C, for penning single fowls. From the shed, a door D' opens into the passage B, and from that a door opens into the roosting room; the perches P P are simple horses, made movable. The nests N are covered so as to protect the hens fouling them, and open by a hanging door into the passage B, so the attendant can gather the eggs without going into the roosting room; a large window at W, and a small one in the rear of the passage, give light and air; an opening for the fowls to pass in and out is made at H. A building of this size, by being made with 4x6 sills, can be moved at will to any part of the yards.

Where it is desired to have all the breeding stock under one roof, and the yards sub-divisions of one general enclosure, the interior of the building may be arranged as shown in Fig. 97. This was designed by the writer for a gentleman in New Jersey.

It is a double-roof house, twenty-four feet wide, and ninety-six feet long. A passage way A A, five feet wide, extends the length of the building, opening into the yards at either end. There are eight apartments for fowls, B B, each 9½x18 feet, with nests N N, arranged as in previous plan to be opened from the passage way. The perches Y Y are movable

trestles, all on a level, and about twenty inches above the floor.
There is a large window in each room as shown at *w w*, and
boles for the egress and ingress of the fowls at *v v*.

The centre of the building, twenty-four feet square, is two
stories high. On the ground floor it is divided into an office *H*,
which may be divided from the passage *a a*, or not, as desired,
and which opens by a door into the yards in front of the building. A room *G*, which can be used as a setting room, the nests
shown at *N N*, and an opening into an out-door run at *v*, (it
was designed for that purpose,) an egg room *E*, with egg cabi-
nets or closets at *e e e*, and a cook room provided with a steamer
or large kettle *k*. *x x x* are spouts to bring the grain and feed
from the feed room above. *D* is the chimney passing straight
up through both stories.

The second floor, (see Fig. 98), is entered by the stairs *S*,
which land in the feed room *F*. *PPP*
are bins for grain, feed, etc. From
this room a door leads into the Incubator room *I*, in which are tables
L L for the Incu- bators. *R* is the
nursery for the ear- ly hatched chicks
—a room 12 x 24 feet, the floor covered with sand, and one large window
twenty feet long, on the east side to
give sunlight. *b b* are brooders for
the young chicks. Other windows are

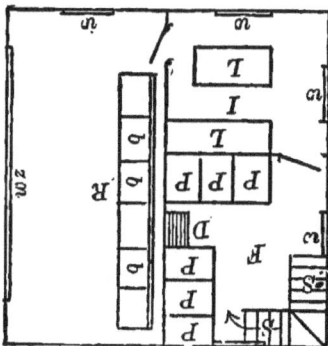

Fig. 98.

shown at *w w* in the different rooms.

If I were to build after this plan, I should re-arrange the second story, so that the nursery would come across the front or
south side, and the Incubator room on the north side of the
house. A room which does not get the noon-day sun is much
the best in which to run any Incubator.

Fig 99 is still another style of building which was designed
for a gentleman living near Philadelphia. It is octagonal in
shape. one story high, and twenty-eight feet in width. The
ground plan is shown in Fig. 100. The door *H* opens into the
passage leading into the central room *M*, which is lighted by
small windows under the eaves of the cupolo. These sashes are
pivoted so as to swing open for ventilation. There are six rooms
for the fowls, and one room *S* which may be used, either for a

setting room or an Incubator room. In case it is preferred to
use the central room M for Incubators, S can be used for fowls,
or a nursery for the chicks.   In this plan the arrangement of the
nests are shown at $N N$, the windows at $W W$, and doors at $D D$;
the nests are facing the wall, and entered at $e e$. The fences
from this building radiate, and allow the yards to be made any
size desired.

Whatever the style of building, ample provision must be made
for ventilation.

In each of the yards, a small shed, not necessarily over two or
three feet high—(a few boards held by a light frame will answer)
—should be provided, under which fix a dusting place of fine

Fig. 99.

sand, wood ashes, and some tobacco dust. It is thus dry in all
weather. In exposed situations, I would make this shed for winter use, by raising the front three feet high, and letting the back
down to and into the ground, having it open toward the south.
In another part of the yard, place a trough or shallow box, in
which keep a supply of fine gravel—unless the soil is gravelly—
in which case it is not needed.

For the young chicks accommodations must be made according to the number intended to be raised. If not over one thous-
and, an ordinary shed facing south or east, with several brooders
of capacity sufficient for that number of chicks will be found all
that is necessary. But if it is intended to go into it on a more

extensive scale, in addition to the buildings for the breeding stock, there will be required an Incubator house or room; a nursery for the young chicks, which should be partly covered with glass, and a second building, into which they should be placed when four or five weeks old. The size of these buildings is to be governed by the extent of the business. Beside these, there might be profitably used one or more sheds, peaked roofed, open on all sides, with perches placed not less than four feet from the ground. After the middle of April, the chicks that are old enough, should be made to roost in these.

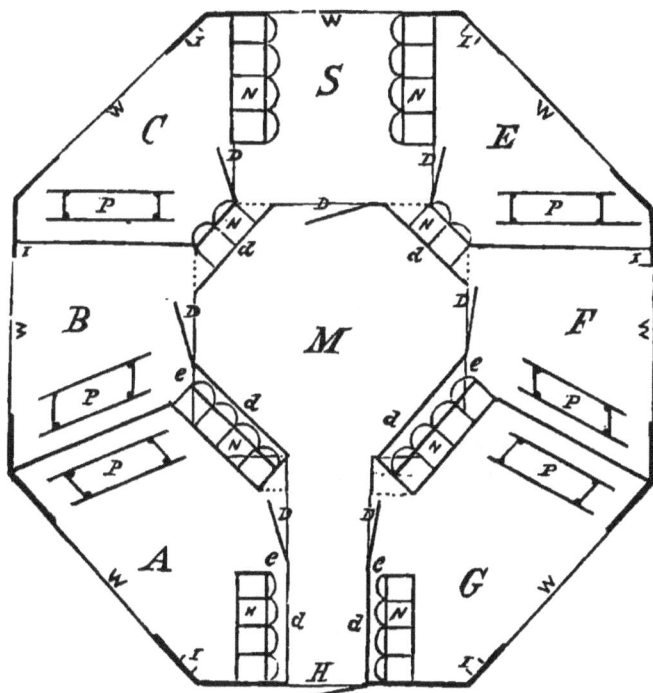

Fig. 100.

A good size for one of these sheds, is six by twelve feet. To build one, set six posts into the ground, three on a side, six feet apart each way; let the posts stand six feet out of the ground. On the top, spike 3x4 joist, all around for the plate; three feet below let in another tier of joist on which to rest the floor, which may be of hemlock boards or any material wished; from the centre of each end erect a stud, on which nail a ridge pole two feet above the plates. Make the roof of tongued and grooved pine boards, battened, letting the eaves project a foot or more.

Lath the sides and ends, or put on wire netting if preferred, from the floor to the roof, making a door two feet wide and three feet high at either end.   Set the perches about two feet above the floor, and as close together as deemed best, perhaps about sixteen to eighteen inches is near enough.   Put a small lath ladder from each door to the ground for the chicks to get up on,   A shed of this size will house nearly three hundred chickens, and when shut in at night they are safe from cats or other nocturnal enemies. The space under the floor affords a dry shelter during rainy days, or it might be enclosed and utilized for brooders and younger chickens.

Fig. 101 illustrates a section of portable fencing made of three upright pieces of 1x2½ inch furring strips, to which is nailed a bottom board ten inches wide and thirteen feet long; three feet ten inches above this another furring strip is let into the three upright pieces, and eighteen inches above that another strip is nailed on top of the uprights.   Full length mason's lath are nailed from the board to the first cross strip, and half length lath above that; this makes a light, neat fence, six feet three inches high.

Fig. 101.

The sections are bound together with old telegraph wire, and are kept upright by braces of the same, fastened to stakes which are driven into the ground until the wire is pulled taut.

In fencing the yards, the height will have to be regulated by the breed of fowls kept.   The Asiatics require only a fence of three feet to keep them within bounds, while the Leghorns, and other light-bodied kinds will readily go over a board or picket fence six or eight feet nigh.   A few years since I put up some fencing of wire-netting five feet wide, with a board underneath, making the fence nearly six feet high, and 1 find the Leghorns are perfectly controlled by it; better, in fact, than by a picket fence, two feet higher, which I had been using.   In putting up wire fencing, never use a top rail; it gives the fowls a foot hold to light upon, and they are certain to fly over.

For the benefit of a class of fanciers who always wish the best of everything, regardless of expense, I append a cut and description of a building built expressly for an Incubator house. See Fig. 102.

The house illustrated is of stone and brick. It may be constructed of wood and answer its purpose equally well. It is sixteen feet square inside; the outer door opens into a little hall or vestibule five feet wide, and extending from the right hand side of the door across to the left wall of the building, and is used for an egg room, having a case of drawers, and a folding shelf under the window on which to place the drawers while sorting the eggs. To the right of the door, in the other corner of the building, is a small dark room in which the eggs are tested by means of a

Fig. 102. AN INCUBATOR HOUSE.

large lamp, a powerful lens, and a large concave reflector. The rest of the building is one room, ten-and-a-half feet by sixteen, and is entered by a door from the hall, right opposite the outer door, and lighted by one window on the north side, opposite the door. The room is provided with tables or stands for Incubators, and a large folding shelf under the window on which to place the egg-trays while they are out of the machines. It has ample ventilation through the cupolo. The building is lathed and plastered, with a dead air space between the inner and outer walls, and also between the wall and the roof. Such a building would have an Incubator capacity of about sixteen hundred eggs; and if of wood, would cost from seventy-five to one hundred and fifty dollars, according to the material and finish.

INDEX: